Nontraditional Careers for Chemists

Nontraditional Careers for Chemists

New Formulas in Chemistry

Lisa M. Balbes

2007

Oxford University Press, Inc., publishes works that further
Oxford University's objective of excellence
in research, scholarship, and education.

Oxford New York
Auckland Cape Town Dar es Salaam Hong Kong Karachi
Kuala Lumpur Madrid Melbourne Mexico City Nairobi
New Delhi Shanghai Taipei Toronto

With offices in
Argentina Austria Brazil Chile Czech Republic France Greece
Guatemala Hungary Italy Japan Poland Portugal Singapore
South Korea Switzerland Thailand Turkey Ukraine Vietnam

Published by Oxford University Press, Inc.
198 Madison Avenue, New York, New York 10016

www.oup.com

Library of Congress Cataloging-in-Publication Data
Balbes, Lisa M.
Nontraditional careers for chemists : new formulas in chemistry / Lisa M. Balbes.
p. cm.
ISBN-13 978-0-19-518366-5; 978-0-19-518367-2 (pbk.)
ISBN 0-19-518366-5; 0-19-518367-3 (pbk.)
1. Chemists—Vocational guidance. I. Title.
QD39.5.B25 2007
540.23—dc22

2006009463

All images appear courtesy of Dana Lipp, Dana Lipp Imaging, 781-380-3851, www.danalipp.us

9 8 7 6 5 4 3 2 1

Printed in the United States of America
on acid-free paper

To the men in my life, Mark, Jack, and Alex
And to my entire family, whose love and support
make anything possible.

Acknowledgments

I WISH TO THANK all the contributors who shared their stories with me and tolerated my relentless probing into their personal histories. If there are errors in my transcriptions of their stories, the fault is mine. Thanks also to all my colleagues and friends who shared stories and resources "off the record" and provided valuable insights that I have tried to weave into these chapters. You're all in here, just between the lines.

Contents

List of Profilees

Sharon Davis Alderman, Weaver, Writer, Teacher, Sharon Alderman Handweaving, Salt Lake City, UT
Alex Andrus, Patent Counsel, Genentech, Inc., San Francisco, CA
Robert Becker, Chemistry Teacher, Kirkwood High School, Kirkwood, MO
Andrew Berks, Senior Patent Agent, Ivax Pharmaceuticals, Northvale, NJ
Olin C. Braids, President, O. C. Braids & Associates, LLC, Tampa, FL
Kerryn A. Brandt, Senior Information Scientist, Rohm and Haas Company, Spring House, PA
David A. Breiner, Manager, Technical Information Center, Cytec Industries, Inc., Stamford, CT
Lucinda F. Buhse, Director, Division of Pharmaceutical Analysis, Food and Drug Administration, St. Louis, MO
Fiona Case, Freelance Journalist, Case Scientific, Essex Junction, VT
Dennis Chamot, Associate Executive Director, Division on Engineering and Physical Sciences, National Research Council, National Academies, Washington, DC
Rani Chohan, TV Producer/Science Writer, National Aeronautics and Space Administration (NASA), Hampton, VA
Veena Chorghade, Consultant and Vice President, Chorghade Enterprises, Natick, MA
John D. Clark, Associate Director, Accelrys, Inc., San Diego, CA
Paul Erskine, Account Manager, Aerotek Scientific, St. Louis, MO
Dan Eustace, Health, Safety, and Environmental Manager, Polaroid Corporation, New Bedford, MA
Donna G. Friedman, Professor and Chair of Chemistry, St. Louis Community College at Florissant Valley, Florissant, MO
Arlene A. Garrison, Assistant Vice President for Research, University of Tennessee, Knoxville, TN
Ted Gast, Managing Director, Carl F. Gast Company, Ladue, MO
Dean Goddette, Director of Business Development, Rigaku Automation, San Diego, CA
Osman F. Güner, Principal, Turquoise Consulting, La Jolla, CA
Ruth Hathaway, Partner, Hathaway Consulting, LLC, Cape Girardeau, MO
Darla Henderson, Senior Editor, John Wiley & Sons, Inc., Hoboken, NJ
Ed Hodgkin, President and Chief Business Officer, BrainCells, Inc., San Diego, CA
David A. Karohl, Director of Business Development, Carbon Nanotechnologies, Inc., Houston, TX

Patricia Kirkwood, Engineering and Mathematics Librarian, University of Arkansas, Fayetteville, AK
Bonnie Lawlor, Executive Director, NFAIS (formerly the National Federation of Abstracting and Information Services), Philadelphia, PA
Thomas Layloff, Principal Program Associate for Pharmaceutical Quality, Management Sciences for Health, Granite City, IL
Joseph M. Leonard, Principal Scientist, Abbott Bioresearch Center, Abbott Laboratories, Holden, MA
Vic Lewchenko, Assistant Project Manager (Contractor), Pfizer, Inc., St. Louis, MO
Dana C. Lipp, Science Photographer, Dana Lipp Imaging, Braintree, MA
John C. Mackin III, Science Department Chair and Chemistry Teacher, Kirkwood R–7 School District, Kirkwood, MO
James G. Martin, Corporate Vice President, Carolinas HealthCare System, Charlotte, NC
Anita L. Meiklejohn, Principal, Fish and Richardson P.C., Boston, MA
Trish Maxson, HR Group Director, Coatings Businesses, Rohm and Haas Company, Philadelphia, PA
W. Val Metanomski, Senior Scientific Information Specialist in Authority, Chemical Abstracts Services, Columbus, OH
Jennifer L. Miller, Executive Director, Corporate Development, Amphora Discovery Corp., Menlo Park, CA
Alison Murray, Conservation Scientist and Associate Professor, Art Conservation Program, Department of Art, Queen's University, Kingston, Ontario, Canada
Lee M. Nagao, Senior Science Advisor, Gardner, Carton, & Douglas, LLP, Washington, DC
Laura M. Rosato, Global Product Regulatory Stewardship Leader, Honeywell Electronic Materials, Honeywell International, Fombell, PA
David Sarro, Freelance Technical Editor, Swiftsure Editing Services, LLC, Cumberland, RI
Sadiq Shah, Director, Western Illinois Entrepreneurship Center, Western Illinois University, Macomb, IL
Joel Shulman, Manager of External Relations and Associate Director, Corporate Research (Retired), The Proctor and Gamble Company; Current position: Adjunct Professor of Chemistry, University of Cincinnati, Cincinnati, OH
Egbert M. van Wezenbeek, Head of Customer Service Development, Elsevier BV, Hoorn, The Netherlands
Nicole Vanop, Senior Director, Customer Services, Tripos, Inc., St. Louis, MO
Alan Gregory Wall, Technical Service Representative, Sigma-Aldrich Corporation, St. Louis, MO
Patricia Hall Ward, President and Founder, Kingston-Ward Advisory Group, Greensboro, NC
James H. Wikel, Chief Technology Officer, Coalesix, Inc., Cambridge, MA
Linda Wraxall, Criminalist Lab Safety Officer, California Department of Justice DNA Laboratory, Richmond, CA

Nontraditional Careers for Chemists

Introduction

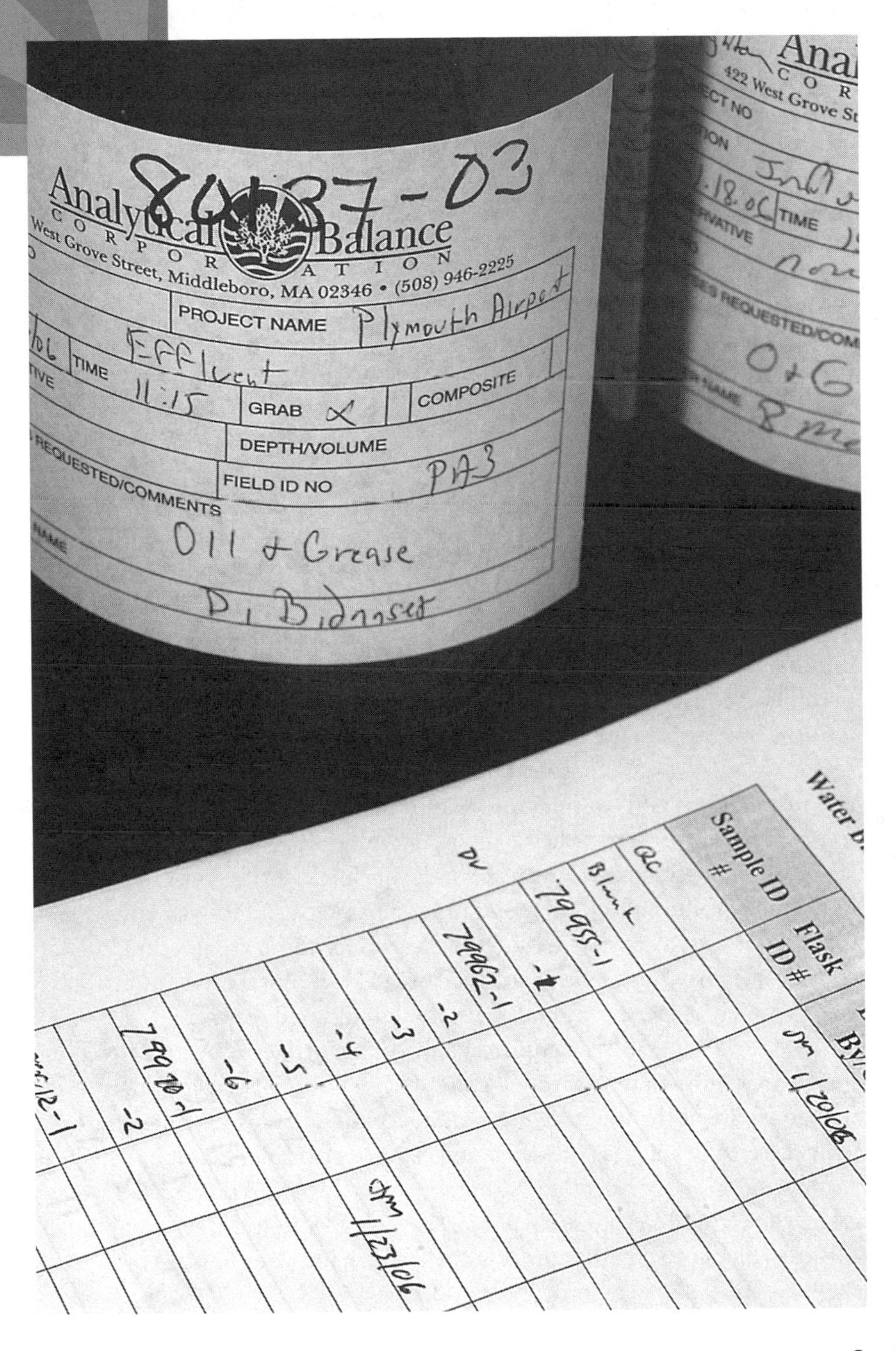

If you've picked up this book, you probably have an interest in chemistry and science and are wondering how you can earn a living doing what you love. Perhaps you have been working as a chemist for a while and are looking for a change or a new challenge. Perhaps you think all chemists are solitary, white-coated geeks in thick glasses, spending their lonely hours staring at test tubes and mixing up smelly chemicals in beakers. While laboratory work is certainly one career path, an interest and education in chemistry opens up a world of professional possibilities, many of which you've probably never heard of, and many of which you've probably never imagined. How about producing video news releases for NASA? Or organizing panels to provide objective scientific advice to the federal government? Or working with scientists to protect and commercialize new technologies?

A background in chemistry prepares you for more than a laboratory career. The research techniques, analysis methods, problem-solving skills, technical communication, and other skills that are all part of a broad science education are of value to a wide variety of employers and are essential for a plethora of positions.

Today, chemists work in all kinds of places, from laboratories to law firms to publishing houses to software companies to art museums. Some continue to do research and expand the boundaries of our collective scientific knowledge; others work in public policy to determine the best way to put that knowledge to practical use and improve people's lives. Some publish and disseminate new information and ideas; some manage the available information and make it easier for other chemists to find what they need. Some create new tools and techniques for other chemists to use; some make sure those tools are applied in an appropriate and safe way. Many spend the majority of their time creatively engaged in collaborations and consultations with other scientists, business people, entrepreneurs, lawmakers, policy analysts, and others.

With all the changes taking place in the business world—the rapid changes in technology, the globalization of trade and markets, and the changing demographics of the workforce itself—it is no wonder there are few "traditional" careers anymore. Nontraditional careers are the rule, not the exception. Chemists are widely dispersed into other fields, and for good reason.

The days of spending an entire career working in one area with a clearly defined career-growth path are long gone. The majority of career opportunities these days are at the interface between disciplines, and as you move through your career, you will use knowledge and expertise gained in one job to move to another position, constantly expanding your skills and knowledge. In many cases, you can add to your skills and stay in the same subject area, or apply your existing abilities to a new knowledge area, to expand your horizons.

If you're thinking about expanding your professional options in science, researching what's out there is the best way to start. Learn as much as possible about what skills and knowledge areas are needed, and in what sorts of environments they are used. By being flexible about not only what you will do but also in what field and for which type of organization you will do it, you vastly increase your employment opportunities and the likelihood of finding enjoyable and lucrative employment that you can be passionate about.

The following chapters are designed to help you start thinking about what career paths you might like to explore further. Each chapter includes a background on the types of positions available in a particular field, followed by detailed profiles of several individuals working in that field. Some planned their transitions carefully, but most were simply willing to learn and were open to the opportunities that presented themselves. By taking advantage of the opportunities that seemed interesting, these people found themselves on a whole new track. In the vast majority of cases they ended up in a career that might never have occurred to them but that they truly enjoy, and now they wish only that they had found out about it earlier.

As you're reading these stories, one thing will become clear. The chapter category distinctions are artificial. Virtually everyone profiled would have fit in at least two different chapters, and in some cases three or more. Their job functions and responsibilities include many different, overlapping areas. In addition, many people moved through several nontraditional careers to get to their current position. Although it can be confusing, the overlapping and blurring of fields and job responsibilities also leads to new opportunities. Regulatory-affairs personnel can move into science policy, environmental efforts flow into standards setting, marketing requires technical communication, and so on.

If there is something you're interested in pursuing, make sure to read (or at least skim) all the chapters that might relate to it. Think about how it could be approached from different directions, and how you could leverage your existing knowledge, skills, and abilities to make yourself valuable in that field. How can the skills you have be applied to a different field, or how could new skills allow you to be employed in a different aspect of your current field?

Another complicating factor is that similar job functions are now being performed by people who come from different fields, and they bring their terminology with them. The same job title in one field can mean something completely different in another field, and the same job function can have different names in different industries. You need to figure out not only what you want to do, and where you want to do it, but what the people in that field call it. Sometimes, just knowing the right words can make all the difference in finding your ideal career.

I hope the background information and personal histories provided here will help you start thinking about all the career possibilities there are and will provide some new directions for your explorations. Just being aware of what's out there will make you better prepared to make choices about your own career path. You never know what's going to happen, and the more you know about the possibilities, the better off you will be.

How do I know? I did it the hard way. After graduate school, I knew I didn't want to be a university professor but didn't know what I did want to be. I took a postdoctoral position, then a "real" job. Eventually, I relocated with my family, only to find out the company for which I was going to work did not get their grant, so I did not have a job. While I was wondering what to do, a colleague called and asked me to help his company with a small project, and said he'd pay me for it. That led to another consulting job for his company, then another client, and so on. After about three years, I said, "Hey, I think I'm a consultant!" Over the 14 (and counting!) years I've been in business, I've learned how to continually acquire new skills, maintain my network, manage a business, market my services, and much more. It's worked out very well for me—I love what I'm doing and have the flexibility to work around my family's schedule most of the time. However, things would have gone much more smoothly if I had known what I was doing ahead of time and had planned and prepared for the transition.

Now you have the chance to do it the right way. Do your research, figure out what you really want to do, and discover how and where you can do it. Your ideal career is out there, but it's up to you to find it. Good luck!

1 Chemistry and Communications

That writer does the most, who gives his reader the most knowledge, and takes from him the least time.
—C. C. Colton

ONE OF THE most highly sought-after skills in the workplace today is communication—the ability to take ideas and data and put them into a concrete format that other people can understand. Technical communicators learn about pieces of a particular subject matter or product until they understand the whole, then they reassemble the pieces in the way that makes the most sense for the intended audience. Scientists do much the same thing—picking apart a particular research area, identifying testable questions, and studying small pieces one at a time until they are able to reassemble their knowledge into a comprehensive understanding of the subject as a whole.

Technical communication (which includes writing, editing, and illustration) involves taking complex information and organizing it into the format that makes the most sense for the intended audience. (Scientific communication is a subset of this field, where the information communicated deals with a scientific field such as medicine, chemistry, biology, geology, and so on.) The information may take the format of a proposal, grant application, users' guide, reference manual, online documentation, slide or video presentation, report, marketing material, peer-reviewed journal article—the list is virtually endless. Technical communicators are employed, for example, by scientific and medical journals, research facilities, government agencies, chemical and pharmaceutical companies, textbook publishers, and nonprofit agencies, as well as in many other institutions.

While almost everyone can write, not everyone can write well. Those who are often asked to review their peers' writings and whose own work is often complimented may want to consider a career in this field. Technical communication is also a good second or backup career, as a significant amount of the work is done by freelancers on a per-project basis and can often be done remotely during nonbusiness hours.

* * *

COMMUNICATING scientific information in an appropriate, informative, and energizing manner can be a fascinating and rewarding career. Scientific communicators provide a bridge between those who create products and information and those who need to know about or use them, and they work in a wide variety of industries and institutions.

Scientific communication (where the subject is scientific) is a subset of the broader field of technical communication, which covers all aspects of taking complex technical information and formatting it appropriately for the intended audience. The final presentation could be a set of Web pages, white papers, scientific reports, online help, illustrations, or something else. The audience may be other scientists in a particular field, other technical people, product purchasers or users (not always the same people), the general public, funding agencies, legislators, or any other group. Communication skills are needed along all parts of the information-creation pathway: from defining a project's scope to researching the content; from choosing a suitable a writing style and creating the content to editing (developmental, substantive, and copy) and indexing it—not to mention the tasks of information architecture and general project management. In some cases, the document will eventually be translated into other languages, so the original document must be prepared with that possibility in mind. If the same information is going to be delivered in more than one format, single sourcing may be used to ensure that changes are carried across all output formats.

A related, but significantly different, field is science writing, a subset of journalism that involves explaining science to the general public, primarily through newspapers, radio, and television. While most professional science journalists start with a journalism degree, some enter with a science background instead. In all types of scientific and technical communication, content is key. Any expressed opinions or feelings must be clearly attributed to a source. They must not be the author's.

Clarity, brevity, and accuracy are crucial when conveying technical information. Science and medical writing is about clearly and concisely explaining technical, and often complicated, concepts to readers. To write effectively about science, a basic understanding of science and the scientific method is required. Ideally, an author will have a thorough understanding of the subject matter to be communicated, but sometimes the author and subject-matter expert (SME) are different people. In the absence of specific knowledge, familiarity with the general subject area and the ability to learn quickly and independently are invaluable. It also helps to be curious about new ideas and to enjoy independently learning new things. Excellent interviewing skills are required to pry missing information out of overworked subject-matter experts.

With technical communication, as with all other technical job these days, computer skills are essential for success. Technical communicators must work with numerous applications in addition to word processing. They must create graphs and charts, chemical structures, and presentations. They must master sophisticated document-delivery systems, portable document format (PDF) documents, Web pages, and much more. In some cases, switching employers will mean learning a new organization's software. Willingness to learn new software quickly, and not let the tool interfere with the product, is essential.

Although virtually every organization needs to produce some kind of documentation, many niche markets go unexplored by those seeking careers as creators of scientific information. For example, some journals hire freelance editors, particularly to assist authors who are not native speakers of English. Creating software documentation and other types of instructions requires logic similar to that used when planning a scientific experiment. Marketing materials, especially for scientific products, must maintain a balance between providing factual information about the product and injecting "marketing spin."

Proposal writing is a skill in and of itself, and those expert in it make an excellent living. In addition to proposals going to the national funding agencies, many companies hire writers for internal proposals. The proposal author must understand the funding policies, procedures, and review cycle, and how funding priorities are set to effectively compete for funding. Being able to identify and respond to requests for proposals and prepare a proposal that will successfully pass the review process is almost an art and requires an author–editor familiar with the entire process.

Moving away from scientific research, all sorts of companies need to explain technical concepts to their users, potential customers, salespeople, and investors. This can include not only product documentation (instructions for use) but white papers, reports, trade journal articles, manufacturing instructions, standard operating procedures, and much more.

Technical communication involves more than just words. While the words are crucial, other aspects, such as illustrations, data tables, graphs, document design, and much more, all contribute to the meaning of a document. In fact, a whole new field called information architecture is now evolving. This field deals with structuring information and data and designing appropriate delivery systems so the intended information is expressed in an appropriate, convenient way for the intended audience. This includes things like determining the best method for content delivery, the interdependencies between various aspects of the data, organizing the flow of content within a particular document, laying out the document, and structuring the style of chapters and volumes.

While there is ongoing debate about whether it is better to start with a degree in communication and learn science in the field, or get a degree in a scientific discipline and learn how to write on the job, there are successful people who have done it both ways. A scientific background will provide credibility with technical colleagues and facilitate communication with them. Writers with a scientific education are often more readily accepted by SMEs than by those with a degree in English, simply because the former are seen as scientists—they're "one of us." However, having a degree in technical communication may make it easier for you to convince SMEs of your expertise in the writing field.

Anyone with a degree in chemistry has spent at least some time writing laboratory reports, journal articles, and so on. This experience provides some insight into what a career in technical communication would be like. However, a scientific education also provides an understanding of the scientific method, the specific vocabulary and rhetorical devices used in scientific communication, the role of peer review, and the research-funding process, and gives one insight into the legal and ethical issues (determination of authorship, conflict of interest, duplicate publication, confidentiality, human and animal research issues). Many of these issues are not addressed in formal technical communication education, so those coming from a science background will have a more intimate knowledge and better understanding of these topics. Those who have been out in the workforce for a while will have significant real-world experience dealing with many of these issues, which can come in handy when moving in a new career direction.

For those who already have technical knowledge, there are many places to acquire or perfect communication skills. Many colleges, universities, and even professional societies offer continuing-education courses in writing and editing. Editing a newsletter, writing trade journal or other articles, or taking on a few writing assignments can be a great way to gain experience and exposure. There are also internship opportunities, some paid and some unpaid, that allow you to "test the waters" to get a feel for the work and see what it's really like.

Technical writers often become technical editors, who are responsible for deciding what content will be covered or included and make higher level decisions about how that material will be presented. For example, editors who work for book publishers may evaluate proposals on the basis of need and projected profitability, decide which projects will be pursued, recruit scientists recognized as experts in their field to write content, and ensure that all the contributions to a given volume are consistent in style. Editors perform a variety of functions, including but not limited to:

- organizing and managing contributions to a multiauthor book
- identifying topics for new books and articles
- identifying potential authors and soliciting copy from them

- improving an author's writing to make it clear and concise
- improving the mechanics of copy (spelling, grammar) and making it conform to company style
- performing fact-checking before publication
- coordinating with reviewers
- designing the layout of the publication
- coordinating with the printer

Often, different types of editors will work on various stages of a particular project—from developmental editors, who work with authors on refining content, to copy editors, who find mistakes in the final text. The same job function may have different names at different companies, and at smaller publications, many of these roles are blended.

In addition to using print media, scientific information is increasingly being communicated visually—using images, video, or both. The general public, and especially young people, expect increasingly sophisticated presentations to engage their interest in science. Producers and consultants with scientific backgrounds are needed to develop these projects. Even Hollywood needs to continually produce movies with better and more realistic science to satisfy a more sophisticated and skeptical public. To address this issue, in 2004 the American Film Institute offered the first-ever Catalyst Workshop, to help scientists and engineers become more knowledgeable about the initiation of motion-picture projects and to teach them how to use their existing skill sets within the film and television profession.

More so than many other fields, technical communication allows for freelancing and makes a good second career, or one that can be pursued while holding a traditional job. Freelancers often work for many organizations, on a project basis. Usually, they have a proven track record and are able to regularly meet deadlines on short notice. An excellent reputation and a good network are essential for success as a freelancer. Furthermore, running a successful freelance writing business requires all the usual small-business skills—setting rates, billing, and continually marketing to identify new clients and projects. Successful freelancers are not always the best writers, but are often the best at managing their business. Most freelancers choose this career for the flexibility, and many work from a home office. Since it is so easy to get started, there is a lot of competition, which tends to drive down rates.

In today's busy world, organizations need people who can communicate complex ideas clearly, quickly, and with the right tone. They must be able to understand the concepts to be presented and to arrange the information in a way that is appropriate for the intended audience. People with these skills will always be needed.

Profiles

Nicole Vanop

Senior Director, Customer Services
Tripos, Inc.

BS, Chemistry, Facultés Universitaires Notre-Dame de la Paix, Namur, Belgium, 1975
Doctorate in Sciences, Chemistry, Facultes Universitaires Notre-Dame de la Paix, Namur, Belgium, 1980

Current Position

Nicole Vanop is currently the senior director of Customer Service at Tripos, a scientific software company. Her responsibilities include documentation, customer support, and software product testing.

Nicole's primary job function is to write more than 35 user guides for her company's various software products. She sees herself primarily as an organizer of information. She talks to the developers to find out what the software will do and how it will work then organizes that into a manual that fits both the corporate style and the users' needs. As the designated writer for several software products, she belongs to many multidisciplinary teams that meet regularly to design and test new functionality. If a new product does not work as expected, the team must determine whether to change the software, the documentation, or the product specifications. Nicole often works on tight deadlines because features are changed up until the end of the development cycle, but the documentation must be ready to ship with the product.

In addition to fulfilling her own writing assignments, Nicole coordinates documentation tasks with the rest of the team to make sure they meet all commitments and deadlines. For example, she creates and maintains a checklist of all documentation tasks to be completed for every release (alpha, beta, final) in the product cycle. When applied to all books that will change in a particular release version, this spreadsheet provides a status-at-a-glance report and allows all team members to know exactly what shape the project is in and where more effort is needed.

As the software approaches "code freeze," after which no additional features will be added, Nicole attends weekly release-coordination meetings with people from the development, customer support, marketing, and quality-assurance teams. Nicole writes tutorials that the software testers then use to check for accuracy and completeness before the product is released. Therefore, the software and its corresponding documentation must be developed in parallel. As director of the documentation team, Nicole is best equipped to coordinate the software testing done by the quality-assurance group and the work of the documentation team with the software-development progress.

As supervisor of the technical support staff, Nicole discusses customer issues with the team manager. She monitors all e-mail messages from customers to the support staff. This allows her to keep up with customer concerns and to identify instances where additional information can be added to the documentation, or even to the software itself, to improve the user's experience with their products.

As department director, Nicole sees herself as a facilitator. Her managerial duties include setting goals for the people who report directly to her and then helping them achieve those goals by coaching them and making sure that they have the tools necessary to succeed. Her own goal is to create and maintain a supportive and enjoyable work environment where team members collaborate with each other and with other departments, making the most of their personal skill levels and individual interests.

Career Path

Nicole became interested in molecular modeling during her senior year as an undergraduate student, while working in an X-ray crystallography lab. Computer-aided molecular modeling was then in its infancy, and exciting new ideas were being published and discussed at international conferences. She chose to pursue a doctorate in that field, in her native Belgium.

At that time, career advancement required a postdoctorate in a foreign country, so she moved to a postdoctoral position at Texas A&M and never looked back.

In May of 1979 she went to St. Louis to visit her college roommate and stopped to chat for a few minutes with her roommate's boss, Garland Marshall. A few months later he called Nicole and offered her a choice of jobs: a postdoctoral position at the medical school or employment in the new company he was about to start. She immediately said yes to the new venture.

Nicole recalls, "Like everyone else I grew up in academia during college and graduate school. From that narrow vantage point, it seemed that academicians were spending a lot of time writing grant proposals and reports, which limited the amount of time they had available to devote to research. I also felt that competition for grant money reduced opportunities for collaboration with other experts. I wanted something different for my own professional career. I am pleased to say that Tripos has fulfilled my expectation by providing an environment where productive collaboration is strongly encouraged and rewarded."

In 1980, Nicole joined Tripos. As the only non-software engineer in a fledgling software company, she earned her keep by writing the user manual. It did not take long before she discovered that she really enjoyed writing. Nicole says, "Organizing the information in a consistent and easily retrievable manner appealed to my sense of discipline. That is still true 25 years later."

Although trained in physical chemistry, Nicole says she "chose scientific writing" as her "professional career." She explains, "More precisely, technical writing found me. I did return to scientific research for about a year, as a Tripos client. But even there my interest and activities quickly turned to documenting software developed by my molecular-modeling colleagues. Having rejoined Tripos, I set up the first customer-training program and was a member of the first technical-support team. This prepared me well for my current responsibilities."

The motto of the Tripos writing team is "ACT," which stands for accurate, complete, and timely. Nicole takes pride in what she does and in ensuring that all three goals are met in every piece of documentation delivered. At Tripos, Nicole says, "There are 3 of us to document the work of 50 developers. That is an extremely high ratio. Documentation work includes participation in project teams, which gives us early and in-depth exposure to new software. We routinely test the software while developing tutorials, and we monitor the defects databases and well as messages sent to the support e-mail alias. We also work on the corporate and technical support Web sites and review marketing documents."

One of the things Nicole likes best about her job is that she gets to interact with many people in other departments. There are, of course, the daily interactions with the writers and support scientists within her own department but also the stimulating work done with software engineers and other members of software project teams. The marketing staff frequently request edits of their brochures, and Nicole works closely with the quality-assurance teams to see that all development paths converge at product release time.

Nicole says, "I am extremely pleased to have been able to draw on my education, experiences, and interests to build a rewarding career. My understanding of our clients' knowledge of chemistry is fundamental to the type of writing I do. As new scientific developments arise and are encoded in our products, I continue to learn new science. I have had—and continue—to learn word- and image-processing applications, so I am never bored."

As for where she will go next, Nicole says, "There is no five-year plan. Opportunities for job diversity abound. All it takes is attention and interest. We have annual performance reviews and goal-setting sessions, with regular updates. Employees are encouraged to develop their own goals, and we do."

Advice

Nicole is fortunate in that most of her time is spent in activities she enjoys. She believes that "recognizing your strengths, and building a career around them, will make you happiest in the long run."

Even though she had been a scientific writer for nearly 15 years, Nicole had never heard of the Society for Technical Communication (STC) until a regional meeting was held in her area. With curiosity and expectation, Nicole checked it out. There she found people with concerns similar to her own—her peers! Over the years, her STC peers have taught Nicole a lot—they have helped her sharpen her tools. Through the journals, chapter meetings, and annual conferences she has gained a deeper understanding of her readers, as well as practical knowledge of various delivery and retrieval mechanisms. Her advice? "No matter what career path you take, find the professional organization that serves it. Join, of course, but also get involved and find out what's really going on in the field."

Predictions

Nicole observes, "The paperless era is definitely here. I noticed that in my own behavior years ago. I print very little and turn to online help when I can't guess things on my own. Our customers have reached that point, too: we have not printed a manual in years. Still, not all online formats are equally effective. I consider PDF useful only for those who prefer to read things on paper. True online format must be presented in a logically organized manner and must make intelligent use of hyperlinks. I have not considered audio or video in the context of my own work. Of course, I have seen it in PC software, but from my point of view, the information must be readily accessible when it is needed, and video is not there yet, at least in my field.

"More importantly, all effort must be made to make the interface as self-explanatory as possible. I champion that concept regularly. Why force the user to wonder where the explanation is when two more words in the dialogue would explain it all?"

Egbert M. van Wezenbeek

Head of Customer Service Development
Elsevier BV

Gymnasium Beta, Pius X College, Beverwijk, The Netherlands, 1982
Drs., Physical and Theoretical Chemistry, Free University Amsterdam, 1987
PhD, Relativistic Density Functional Theory, Free University Amsterdam, 1992

Current Position

Egbert M. van Wezenbeek is currently the head of Customer Service Development at Elsevier BV. His team consists of a group of project managers, a marketing communications manager, and a reporting and metrics unit.

The Customer Service Development (CSD) group consists of 11 people and is responsible for developing tools and services aimed at the editors, authors, and reviewers associated with the journals published by Elsevier. Egbert's unit is responsible for coordinating communications to all these groups and for developing performance metrics for the journals. Journals are critically dependent on good and professional services being delivered to editors, authors, and reviewers. Services can include electronic submission of manuscripts, coordination of peer-reviews, copyediting, typesetting, and so on. Egbert's team is involved in all the above-mentioned tasks.

Egbert's main project is the rollout of a Web-based submission-and-review system for all 2000 + Elsevier journals. In previous positions at Elsevier, Egbert was part of a project team that set policies, defined internal procedures, coordinated the rollout within his own department (which covered 90 journals), and ensured support for all journal editors during the rollout phase. Even though this system has been in use for several years, it is essential to continue to monitor editors' comments to identify problems and improvements to address in future releases.

For certain subject areas, Egbert knows the important people and developments taking place and maintains connections with those people. This includes occasionally meeting journal editors at conferences.

Egbert conducts most of his business by electronic mail. Each message is either answered or forwarded to one of his team members for answering. An essential part of his working day is having progress discussions with team members and having group meetings related to existing and proposed new projects. Egbert usually checks e-mail again in the evening—both to give himself a fresh start in the morning and to avoid losing an entire day when corresponding with colleagues in the United States. Usually there are multiple projects in progress simultaneously, so setting priorities and having good time-management skills are very important.

On a typical day, Egbert receives messages from all parts of the Elsevier organization and sometimes also from journal editors, authors, and reviewers. These messages usually come indirectly through colleagues, when there are questions on how to use the Web-based submission system. Journal editors and article authors often have questions about special issues of upcoming journals. In some cases, they send in suggestions for new features or enhancements that would make the system better.

Over the years, Egbert has built an extensive network of contacts with all parts of the Elsevier organization, ranging from publishing, production, and marketing, to finance and customer support. These contacts are essential in his job: they help him get feedback on project progress, obtain a critical mass of support for getting new projects established, and, perhaps most importantly, they help him gather ideas for improvements and new projects.

Travel, and preparation for it, also take up time. Egbert mainly travels to meet Elsevier colleagues around the world. He is currently on the road about 10% of the time, down from 15–20% when he was still building his network of contacts and working in primary publishing. Though travel itself is not a problem, being away from his family is sometimes difficult and requires that he plan his trips so that family life is interrupted as little as possible. An understanding family is also a valuable asset.

Career Path

During his doctoral studies, Egbert's work was mainly theoretical, and by the end of his graduate career he was ready to do something more practical. He was intrigued by the publishing process, having authored several papers, and thought he might be more successful in an industrial position than in an academic one. He also enjoyed and wanted to maintain the many international contacts he had made while a graduate student. After giving it some thought, Egbert realized that working for a publishing company would be a good fit for his career goals. His sister was working as a desk editor at Elsevier, so Egbert applied there and was hired as a junior publishing editor. They liked his analytical and communication skills, his interest in the field of publishing, and the fact that he had been involved in scientific research.

In his first year at Elsevier, Egbert was given all kinds of small projects in the chemistry department, to familiarize him with the work of a publishing editor and allow him to learn how journals and books "operate" without having the actual responsibility of running a journal or book project. This was similar to a traineeship and was excellent on-the-job training.

At the end of the year, he became a junior publishing editor in the chemistry department. He continued to move up over the years, becoming a publishing editor in 1994, then moving to the department of physics and the position of senior publishing editor. Egbert became a publisher in 2000, then moved back to the chemistry department in 2001 to become a publisher of physical and theoretical chemistry.

As he moved up, Egbert had "an increasing responsibility for the portfolio of journals and books handled, in terms of numbers and decision power, plus the addition of management tasks." He explains, "As publishing editor one handles journals and books, starting as a junior publishing editor in the first year with a smaller portfolio. At Elsevier, a team of publisher and publishing editor handles a subject area (for example, physical and theoretical chemistry) with clear task definitions. The publisher is responsible for the strategic direction and knowing the important people in the field (so has to network intensively at professional society meetings and through campus visits) while the publishing editor is responsible for the day-to-day management of individual journals and the execution of the strategic plans as set by the publisher. Good cooperation between publisher and publishing editor is essential to ensure journals run well. Usually people start as a publishing editor, then progress toward senior publishing editor if they have shown high skills in the management and development of journals. Development to publisher is the next step."

For Egbert, the progression toward publisher was due to his having completed important strategic initiatives that were related to his portfolio of journals and books. His promotion was the result of providing benefits to the company. For example, early on Egbert noticed that many other companies were starting to put portal Web sites on the Internet. He proposed, and got Elsevier to sponsor, a Web portal for the surface-science area, which launched in 1999 and was very successful. In fact, competitors used it as an example when designing their own portal Web sites. Being involved from inception to end in a successful, company-wide project helped cement Egbert's reputation within his company.

Working successfully with people from many departments, and at many levels, also helped broaden his understanding of how the company worked and what he needed to do to be successful within that culture.

Egbert interacted daily with scientists in the broadest ways. He says, "They publish in our journals, review articles for us, and are the editors of our journals. Some editors are special due to their high esteem and expertise in their field, and it is a privilege to know them and be able to discuss science and politics with them. My contacts appreciated being able to talk about their science with me since I have a chemistry background. This helped considerably. Also, the fact that I had worked with computers for a very long time has been useful, since information technology is becoming more and more important."

Being able to understand and empathize with his customers' issues allowed Egbert to more fully comprehend their problems and concerns. He was then better able to find ways to solve their problems. "Creativity in finding solutions to problems is also important," he says. While the contact with scientists was inspiring, Egbert also enjoyed the travel and the chance to experience many different places, people, and cultures.

In 2006 Egbert assumed the position of head of Customer Service Development, where he will apply his skills in a different way, developing tools for editors, authors, and reviewers. His many years working in primary publishing have given him "the background needed to know what the individual scientists expect in terms of systems and services."

As for specific skills needed to be successful in publishing, Egbert feels that analytical skills, organizational skills, problem-solving skills, and the ability to be tactful and diplomatic are all important. "You must be able to represent, and sometimes to defend, the company and its decisions. Also, being friendly to everyone, no matter what their attitude, is essential, as is not letting the little things get to you." Egbert feels that the analytical skills he developed during his graduate study are still valuable assets. He notes, "Studying chemistry teaches one to be organized and to use effective analytical strategies when solving problems." Most scientific publishers prefer to hire people with a scientific background for jobs related to primary publishing, since they will be better able to relate to the scientists who are their customers.

Egbert's position is especially challenging since science is developing so quickly, and the pace seems to be increasing. Furthermore, publishers like Elsevier must deal with business issues such as converting the traditional subscription–pricing structure to one based actual usage of journal articles; dealing with open access and related "information should be free" initiatives; continuing to offer good products and services; continuing to increase the value of the publisher; and constantly reevaluating the entire publishing process.

However, the opportunity to translate new developments in science into new publications and to develop new tools and systems to assist editors, authors, and reviewers makes it worthwhile. It does not hurt that publishers' salaries are comparable to, or perhaps a bit higher than, those of laboratory chemists.

Advice

Egbert advises those interested in this field to "try to learn as much as possible about the publishing process and the companies that are currently active in the field." He says, "Develop good analytical and project management skills, and be able to point to specific examples of these in your work history. Having some knowledge about science in general and information technology will also come in handy. Of course, expert knowledge about the publication process and market situation would be ideal, but most publishers will train new employees."

Predictions

Egbert predicts "that publishers will remain an essential partner in the publication process, though the exact relationship will change." He adds, "Already things are very different from just five or ten years ago! Accessibility and retrievability will become more and more important, and the printed journal might disappear in another five to ten years. The printed journal as a brand identity will remain, and journals as a concept will not disappear, since they constitute the collection of articles in the fields they cover. New journals will continue to be started, since science always develops and has done so ever since the start of the scientific journal in the 17th century. Poor journals need to be discontinued."

David Sarro

Freelance Scientific Editor
Swiftsure Editing Services, LLC

BA, Philosophy and Biology, Providence College, 1983
Associate's degree, Chemical Technology, Community College of Rhode Island, 1988
MA, Professional Writing, University of Massachusetts, Dartmouth, 2003

Current Position

David Sarro is a freelance scientific editor, running his own small business. As a technical editor, David feels his job is to "make the proverbial sow's ear into a silken purse," which requires deep thought and creativity. However, being an editor instead of an author allows him to avoid dependency on subject-matter experts (SMEs) to feed him technical information, which enables him to work more independently. Editing is essentially a running and thoughtful commentary on the writer's work, carried out in the margins via emendations and suggestions.

David says, "When I worked as a food-service manager, I noted a curious phenomenon. In casual conversations with customers, I realized that many people know little about food other than what they themselves liked. Time after time they would imply that the mere act of eating conferred general knowledge about food, particularly about its preparation. As an editor, I've observed a curious parallel: many people who write, write poorly. Yet their attitudes suggest they believe that just because they write, they write well. The most difficult part of my job is not the technical part of emending the copy in front of me. It's trying to tactfully suggest to writers that they're not Samuel Johnson."

Career Path

David took a long and winding road to arrive at his current situation. In his early years, he worked as a vending-machine repair technician for a large industrial food-service company, where he received promotions to operations supervisor and then to operations manager. That latter job taught him the usual business skills, but while there he realized that "ideologically speaking, I'd be better suited to empty spittoons in a tuberculosis ward than to advocate corporate interests as a management drone."

David left the food-service company in 1981 and returned to college full-time. He pursued a double major (philosophy and biology), earning his B.A. in 1983. Upon graduation, he started law school, and upon completion of the first year he began an internship with the Rhode Island attorney general's office.

However, for financial reasons, David was forced to leave school and return to his old career of refrigeration mechanic.

He was not satisfied with (as he then saw it) having taken two steps backward. He was ambitious and attended the Community College of Rhode Island part-time, earning an associate's degree in chemical technology. Upon graduation in 1988, DuPont hired him as a senior technologist in the company's now-defunct Medical Products division. There, David followed established synthetic protocols, working with ^{3}H, ^{14}C, and ^{32}P, synthesizing the company's catalogue offerings. Simultaneously, he continued his chemistry education, matriculating in several upper-level chemistry and calculus courses at Rhode Island College and Northeastern University. Eventually, he was promoted to chemist at DuPont.

David recalls, "Oddly enough, my experience at the chemistry bench introduced me to technical writing. The synthetic protocols I followed as a production chemist were poorly written. They were ambiguous and unclear, and I would routinely take it upon myself to rewrite them before I used them. After a while—and to my considerable surprise—I realized I actually enjoyed the task. I had always been a wordsmith of sorts, and I welcomed the challenge of expressing information so that it would be both readable and precise. Serendipitously, DuPont at this time had applied for the International Organization for Standardization's IS0 9001 accreditation, which meant it faced the massive task of formalizing its operational documentation, including its protocols. Inasmuch as I had already been rewriting documents for several years, it was no surprise when management assigned me the task of revising the bulk of its protocols to conform to ISO guidelines."

David thoroughly enjoyed his writing responsibilities—so much so that he decided to make technical writing his career. He entered the professional writing program at the University of Massachusetts, Dartmouth. At about the same time, after some two-and-a-half years of intense preparation, DuPont obtained its ISO accreditation. However, this meant that David's writing tasks officially ended, and he found the prospect of returning to the radioactive bench full-time unsavory. So, he sought work elsewhere as a technical writer.

David soon identified an opportunity at a relatively small company that developed software for the shipping industry. Armed with almost three years of technical writing experience and two part-time years of graduate study, he started work. Unfortunately, he found the subject matter uninteresting and the work environment uninspiring.

In the fall of 2000, David moved to the technical communications department of Waters Corporation in Milford, Massachusetts. He was recommended by a coworker from a former job and, as an ex-chemist, was hired almost immediately.

Since his professional interest lay in producing rhetorically effective documents, David naturally gravitated toward editing and soon became the department's assistant technical editor. When the senior technical editor was laid off, about three years later, David assumed her responsibilities, and soon thereafter he satisfied the requirements for his master's degree.

While David enjoyed working for Waters more than for any other employer, he still felt that "corporations benefit far more from the efforts of conscientious and energetic workers than the workers themselves do. Therefore, to the extent possible, people should own their own businesses so that their strenuous efforts—the long days, frequent weekends, and exhaustive mental exertions—fully redound to their own benefit." To this end, David established his own business, a scientific editing service, with Waters as his primary client. He now sets his own hours and rates, and his rewards are commensurate with his efforts.

The most difficult part of David's job is not the editing itself but tactfully presenting suggestions for improvement to authors who may not be receptive. Of course, writing abilities and sensitivities of authors vary widely. For example, David says, "The technical writers in Waters' technical communications department usually show more writing skill than the engineers, scientists, and software developers whose journal submissions and white papers I edit. Nevertheless, many are fairly thin-skinned and tend to receive positive criticism as unwarranted, or, worse a personal slight. To counter this effect, I remind my clients that all writers err, that even professional editors need editors, and that I fully expect to encounter many errors in their work. Moreover, I suggest—as subtly as possible, mind you—that my only interest is to help improve their writing and that I'm not interested in judging their abilities as writers or demonstrating superior knowledge of the discipline."

David's success at Waters, and now in his own company, is the result of many factors. His chemistry background made him familiar with many of the company's products, so he didn't have to struggle to assimilate a new, alien technology. He was able to combine his experience with a strong interest in writing effectively and to perform at a high level from the very beginning.

When David decided to focus on editing, his chemistry background again helped him understand the subject matter. Beyond that, the analytical rigor of science and mathematics training and practice helped him work efficiently through editing problems, particularly those involving a document's overall organization and, at the sentence level, syntax.

David says, "Given the chance to relive my life, I would want to become an opera conductor. Failing that, I would want to be an operatic tenor, a philosophy professor, or a physician, in that order. I wouldn't, however, aspire to be an editor. But that would require a depth of self-knowledge that I simply did not have this time around. Nevertheless, I'm fairly content in this profession, and I expect to stay—at least for a little while."

Advice

David notes, "Technical editing—and editing in general—is like any other profession: invariably, the most effective practitioners are those who have both natural predilection and interest. So the most important trait for anyone who wants to become a technical editor is a love of language. He or she must possess a perceptive 'inner ear,' a 'feel' for the subtleties of phrasing and the nuances of meaning. In my own case, having studied foreign languages—French, German, Italian, and particularly Latin—has helped me tremendously because it developed the kind of sensitivity to language that I, as an editor, find indispensable.

"Yet technical editors must be more than merely intuitive. They must also be expert at analytical reasoning. The ability to deconstruct a document, chapter, paragraph, or sentence to understand what a sometimes-confused author is trying to express is crucial to editorial success. Thus, editors must know cold the rules of grammar and how to apply them. Likewise, they must know which stylistic conventions to enforce in the document before them.

"As important as analytical reasoning, in my view, is the need for technical editors to be conversant in the subject matter they're editing. I've seen several instances where editors who have the requisite 'feel' for language and absolute mastery of style, syntax, and grammar have failed because they simply didn't understand the subject matter of the documents they worked on. For example, I'm largely ignorant of banking operations, so I would be foolish to accept as an editing assignment a highly technical document describing loan financing. Trying to edit content you don't understand can be a humbling experience, and one would do well to know his or her limits."

Predictions

"As for the future of technical editing, I think it's promising. In the current, rather lackluster job market, there are nevertheless many advertised positions for editors. The reason is obvious. Frankly, most technical writers—at least those I've encountered—write poorly. So organizations who care about the quality of their documents will always engage an editor. Moreover, as the phenomenon of 'off-shoring' continues to gain momentum within the technical-writing industry, I expect that editors in the United States will be even more in demand. I've seen some pretty hackneyed prose emanating from English-speaking places like Singapore and India. Such documents absolutely need rescuing by an editor if they are going to be useful to an American audience."

Rani Chohan

TV Producer/Science Writer
National Aeronautics and Space Administration (NASA)

MA, Journalism, University of Wisconsin–Madison, 2002
BS, Chemistry, Elmhurst College, 1996

Current Position

Rani Chohan is a science writer and TV producer for NASA. Her job is to communicate the science of NASA to the rest of the world by creating stories and pictures that tell a compelling science story to the public. She also produces content for their Web site, www.nasa.gov. In most cases, her productions are aimed at the general public, so the science must be explained at a very basic level.

Rani monitors the latest news, both from NASA and from outside sources, and regularly talks to NASA scientists about their cutting-edge research. At media strategy meetings, Rani and coworkers decide which stories to pursue, which to put on hold or not pursue, and how to respond to stories about NASA that have been publicized by other organizations. Of the stories they decide to pursue, they decide which ones will appear on the Web site and which will be released by other methods. They decide which will be produced on video, and which will be text accompanied by still images. Once a story has been set in motion, Rani works with animators and visualizers to produce images that support the story. She normally produces content for the Web site and so designs a series of Web pages that tell a compelling story.

Rani is also responsible for talking to reporters from other organizations and explaining NASA's research and results to them individually.

Career Path

For as long as she can remember, Rani wanted to work for NASA. She ended up there, and in her ideal job, but not by taking a predictable path.

Rani graduated from college with a degree in chemistry. She recalls, "I wish I was a college graduate who knew exactly what they wanted to do when they finished school . . . but I wasn't. I spent the first couple of years exploring careers and trying new things." She spent some time as a quality-control chemist at Blistex but found herself "very bored." She says, "After all the hard work to get a degree in chemistry, this job did not even require me to think!"

Rani then got involved with several political campaigns and for a few months worked with TV and the media as a full-time volunteer. This was an eye-opening experience for her and convinced her that she wanted a career in communications or public affairs. However, she had absolutely no idea how to go about it.

Rani tried to get a position in public affairs but was unsuccessful. She went to a community college to take some classes in business, hoping this would open doors. It did, and she ended up getting a job in sales, which she soon learned she did not enjoy. However, the experience taught her some very valuable skills, some of which she still uses when "selling" stories to reporters.

Then, Rani stumbled into broadcast news. She volunteered to work at a small TV news station in Rockford, Illinois, on weekends, while continuing the sales job in Chicago. Rani recalls, "Everyone assumes you're smart if you have a chemistry degree. It is extremely surprising how many doors open when you tell a prospective employer or graduate program you have a degree in chemistry."

After about three months of volunteering, Rani was offered a part-time job as a weekend TV producer at $8 per hour. As a producer, she was the "captain of thenewscast." "I decided what goes, assigned stories to reporters, picked which video to use and wrote the show."

Rani worked in Rockford, Illinois, for 10 months, then moved to a Madison, Wisconsin, television station. News is an extremely challenging field, and Rani was never bored. She also learned to become detailed-oriented, as she alone was responsible for all aspects of the production. She was a morning producer, which meant she worked overnight for two years. The job didn't pay much, so she had to get a second, and then a third, job to make ends meet. She wanted to take her writing skills to the next level, so she began taking classes in journalism at the University of Wisconsin.

After working overnight for two years while going to school part-time, Rani suffered an episode of depression, brought on by a lack of sleep. She had a tough time improving her writing and broadcasting skills, and her health suffered. Rani discovered there was a journalism program focused on science writing. She wanted to apply, but her undergraduate GPA was not high enough to meet the program's entrance requirements. However, since her degree was in a hard science and she had related work experience, they allowed her to enroll as a full-time special/probation student in the graduate journalism program.

As part of her probation, Rani was required to earn good grades during her first semester to remain in the program, which she did with ease. "After P-chem, it's amazing how easy other classes were. I got straight A's . . . even in some of the very competitive classes."

Rani's graduate adviser was a science-writing Pulitzer journalist, president of the National Association of Science Writers (NASW), and a wonderful writing teacher. She helped Rani improve her writing skills and gain confidence in them. Her professor also had a lot of professional connections and was able to introduce Rani to a science writer who worked at NASA. Rani interviewed for an internship, but the job went to one of her good friends.

Shortly thereafter, a position as a TV producer for NASA opened up, and the friend immediately thought of Rani and told her this was "her job." At the time, Rani was tired of broadcast TV and really wanted to be a print writer. While attending her second NASW meeting, one of her professional contacts showed her the NASA job listing and said that she should apply for it. Still, Rani hesitated. Then her brother, a reporter for CBS, called and said one of his friends forward him a NASA job listing, and he thought Rani should apply. After three completely different people told her the job would be perfect for her, Rani gave in and applied. She was hired and has been happy ever since. Rani feels she was right for the NASA position because she had an undergraduate degree in science, because she gained experience in broadcast news (working for free or near-free), and she had a graduate degree in science journalism.

On a typical day, Rani may work with an animator to create animation that explains a scientific concept, escort TV or movie crews, answer media inquiries, help TV and film producers obtain NASA video for documentaries, produce video news releases (a press release in video format) to accompany science results, write articles about the science being done at NASA, work on producing a press conference (which can take weeks or months to set up), coordinate an interview between scientists and reporters, and much more.

Rani works on breaking news stories, such as fires and storms, and obtains satellite images to send out to the press. She is always investigating potential stories and interviewing top scientists about their research and says, "I'm always learning something new. I feel like I'm in school every day."

Rani was away from chemistry for five years before she was able to use it again. While specific technical knowledge (like spectroscopy) has been extremely useful, having the background gives Rani a context for discussions and makes it easier for her to communicate with scientists. She also knows where to start looking when she doesn't understand something.

After working with a particular science topic for a long time, Rani becomes an expert in that field also. She's met some extremely interesting people, including astronauts, actors, directors, and politicians.

Rani loves the creativity of her job and the flexibility of her schedule. The worst part of her job is the politics. She notes, "You'd be surprised at how much science is controversial and receives politic scrutiny. Who ever thought the weather could be political?" Scientists are often afraid of how their peer group will perceive them if they seek media attention. For these reasons, many scientists shy away from public outreach and media attention. "It is my responsibility to recognize that fear, and also to realize that not everyone can be put on camera, no matter how important their science is or how great a scientist they are," Rani says.

For the future, Rani is interested in moving up at NASA, but she is holding back a little until she has children. She says, "As soon as my kids are at the right age, I plan to take on a more vigorous career path." She would like to stay at NASA as long as possible and is looking forward to "working with scientists who are developing plans for sending people back to the moon, Mars, and beyond." She adds, "The science is very exciting—even as we learn about our own planet." Rani also wants to grow her communications skills and develop her interests in politics by working in legislative public affairs or perhaps science policy. She has thought about becoming a press secretary. "Luckily, all these interests are related," Rani observes, "and I will take advantage of whatever opportunities come along."

The financial aspects of working in science films are interesting. Especially in television, most of the training is on the job, "You learn by doing," Rani points out. When first starting out in film, a lot of people work for free, but that shouldn't continue for very long. Once you have some experience, the potential to earn money is endless. Rani says, "It feels like I make way too much money for what I do, because I'm having so much fun. How much you can make depends where you work. If you stick with it and do what you love, the money will come."

Rani continues to learn and develop professionally outside her job, helping out on independent films and looking into producing science documentaries.

She firmly believes that the skills that have made her successful came from her study of chemistry, "especially being able to take a large, complex problem and break it down into solvable parts."

Advice

Rani believes that "in order to work with a science organization, the biggest requirement is a love of science. You must like to read and learn about science." She says, "You don't necessarily need a science degree, but it does come in handy. And it doesn't hurt to be an extrovert.

"Explore. Don't just major in journalism or science. Make sure you have a well-rounded and diverse educational background. Take your writing and communication skills seriously. NASA and many other scientific organizations are desperate for scientists with good communication skills. One good way to practice is try to explain your research to your grandmother or neighbor. If they understand and engage in the conversation, then you know you're a pretty good communicator."

Knowing the media and having a strong journalism background is also very important. Rani advises, "Spend some time working at a newspaper or TV station before going to work in any type of public affairs job. The experience will give you an edge and make your resume stand out from the rest. Make sure you get a good internship somewhere. The experience and contacts will be invaluable in obtaining future employment.

"Develop and maintain a writing portfolio, and get published, so you have specific examples to show to prospective employers."

Predictions

Rani says, "I think there is a lot of opportunity in science writing and producing. *Discovery, National Geographic, NOVA,* and many other publications have science programs and need materials, as well as journals, magazines, newspapers, pamphlets, and newsletters. The *New York Times*'s *Science Times* is probably one of the most famous newspaper science sections . . . although I'll admit that newspaper science writing sections are struggling across the country. However, health stories, especially women's health, are always hot topics. Every science or technological organization needs good science communicators."

Fiona Case

Freelance Journalist
Case Scientific

B. Sc. (Honors), MRSC, Chartered Chemist, Birmingham University, England, 1986

Current Position

Fiona Case currently works as a freelance scientific journalist. Most recently, she has been writing feature articles for *Inform* magazine (AOCS Press) and *Chemistry World* (Royal Society of Chemistry). Writing is one of several activities that occupy Fiona's time. Fiona has a "portfolio career," meaning she has several different income-producing activities. She writes; is a consultant for companies in the personal care, cosmetics, and food industries; works with the Nano Science and Technology Institute as a lecturer and organizer for the Nanotech conferences; and is involved in a group writing new software for computational chemistry.

Fiona works from a home office. During a typical day, she will answer e-mails, interview people for articles, write queries, and write some articles. About 20% of her time in her home office is spent reading—studying the literature and patents, and the work of other science journalists. Fiona feels that reading articles written by other science journalists is one of the best ways to learn how to write.

She also spends a significant amount of time traveling, attending conferences, and looking for new ideas that could be turned into queries. A query is an idea for an article, which is sent to the editor of a particular publication. It is usually accompanied by a brief summary of the idea, a letter explaining what expertise the author has to write the article and why the readers of the publication will want to read it, and clips of previous articles the author has written. If an editor accepts a query, the freelance writer then receives a contract to write the proposed article at the publication's prevailing rate, by a specific deadline. Successful queries involve understanding not only the types of articles the publication prints but also the readership demographics and history of coverage of that topic.

Career Path

Fiona began her career as an industrial polymer scientist in the late 1980s at Courtaulds Research in the United Kingdom, working with solvent-spun cellulose fibers and films. She was also involved in early efforts to use computational chemistry in an industrial environment, which is what brought her to the United States in 1991 to join Biosym (then a small start-up computational chemistry company in San Diego).

Fiona spent eight years at Biosym (now Accelrys), conducting contract research for some of the top U.S. and European materials-science companies to demonstrate the value of molecular modeling for their specific applications. As manager of their materials-modeling training group, she prepared and presented workshops worldwide on the molecular modeling of materials and polymers. She also worked in marketing and technical sales support where she learned valuable business skills. Fiona says, "If you are flexible and enjoy new challenges, I would definitely recommend working for a small start-up company early in your career. It may not pay well, and the hours are very long, but you get the opportunity to do so many different things. It's a wonderful learning experience. I even starred in a promotional video!"

In 1999, Fiona moved into a central research group at Colgate Palmolive. This presented new challenges, including materials structure and property prediction for toothpaste, detergent, hard-surface-care and personal-care products, and packaging and fragrance technology. She continued to be active in professional development while at Colgate: she taught short courses for the American Chemical Society and organized a symposium on Mesoscale Phenomena in Fluid Systems, which led to her becoming a coeditor of a book based on the symposium.

Eventually, Fiona decided to create a career that would allow her to undertake a wide range of activities. After more than 15 years working for various companies, large and small, she decided to start working for herself.

The most common way to enter a scientific journalism career is probably to take a course on the subject, and the National Association of Science Writers (NASW) provides helpful advice on this subject. Fiona did not take a formal course but did read several publications, including the NASW *Field Guide for Science Writers, The Associated Press Style Book, Eats, Shoots and Leaves,* and several other writing books to polish her writing skills. The most important skill in a freelance writer is, of course, the ability to write well. A broad knowledge of chemistry, the ability (and the desire) to rapidly learn new ideas, and the confidence to go up and talk to strangers are also required. Expertise in literature searching, and the habit of regularly reading technical journals are also helpful.

Fiona started out by volunteering to write conference reports and book reviews for a trade journal, doing so during evenings and weekends while still employed in a traditional job. She was fortunate to have an editor who was willing to provide guidance and advice. While some editors are willing to offer guidance and advice, very few of them, however friendly, have time to correct basic errors of grammar or style.

Fiona believes that chemistry should be reported in an accurate and interesting manner, which is what she aims to do. Although she currently writes mainly for an audience of chemists, she is expanding her skills and hopes in the future to also write about chemistry for broader public consumption. Fiona notes, "Once you are working regularly with a particular publication, your editor may ask you to write on a range of topics. I love that challenge." Fiona enjoys what she does, and adds, "Writing gives me an excuse, and in some cases financial support, to attend scientific conferences. I go to seven or eight each year. It also gives me the impetus to learn about new areas of chemistry."

Freelance writers do not receive a salary but instead are paid for each piece of work completed. They are usually paid upon publication, which can be several months after submission of the final draft. As writers become more experienced, they can command higher rates. Freelancers are fairly well paid, especially if they can write fast, but Fiona "wouldn't expect to become rich with this job—although some people do." Freelance writers are running a business and need to be concerned with all the business aspects as well—marketing, invoicing, setting rates, and so on. In Fiona's case, she created a Web site for her company. She doesn't really expect potential clients to find her via the Web site, but it is useful to be able to refer potential clients to her online biography or list of clips.

Fiona has now been working as a freelancer for several years and has no plans to change anytime soon. She says, "One nice thing about working for yourself, and about freelance journalism, is that you can live wherever you like. I live in Vermont, a beautiful part of the world."

Advice

Fiona cautions, "The number one rule of freelance writing is 'always complete your articles on time.' If you do not, you likely will never work for that editor again." Regarding getting started, Fiona suggests taking a course in journalism if you have the time and money. She advises, "If that is not possible, start writing now, in evenings and weekends, while still employed in another position. Most local newspapers are happy to receive well-written articles (especially from scientists who are not expecting to get paid). Alternatively, volunteer to write book reviews and meeting reports for trade journals, or start writing for an online journal. Even though you won't get paid, you will then have clippings, which will be valuable when you start to approach editors.

"Be active in your professional organizations, such as the American Chemical Society, the American Oil Chemists Society, and the Royal Society of Chemistry. If you participate in professional and outreach activities you will build a reputation, and you will make contacts." Fiona particularly recommends offering to organize symposia for national and international meetings as a way to meet leading scientists in your field and expand your network.

Darla Henderson

Senior Editor, Chemistry and Biochemistry
Manager, Product Development, Databases
John Wiley & Sons, Inc.

BS, Chemistry, Appalachian State University, 1994
PhD, Bio-organic Chemistry, Duke University, 1998

Current Position

Darla Henderson is a senior acquisitions editor at John Wiley & Sons, Inc. As such, her primary responsibility is to identify the information needs of chemists and biochemists and to work to locate the best authors and editors to create projects that meet those information needs. To do this properly, she has to be aware of the important developments in the field, which involves keeping up with the research literature, attending symposia and conventions, and talking to authors and others in the community. Additionally, her responsibilities include managing the review process for new projects, developing and refining the business model for all projects, compiling basic project information for marketing purposes, identifying marketing opportunities for each project, and ensuring that the project meets the desired information needs upon completion.

Darla also manages the financial aspect of the chemistry and biochemistry product line, including planning and budgeting for revenues and expected costs, and training and managing other editors and assistants in the editorial program. In some cases, the work is conducted in collaboration with a society or a corporation, and this requires negotiation for acquisition of a product or a joint copublishing arrangement, such as with Organic Reactions, Inc.; Organic Syntheses, Inc.; or the American Ceramic Society.

Darla also manages product development for database products that have functionality such as structure, reaction, or property searching. She is currently working with colleagues in the higher education division to implement an online version of the *Textbook of Biochemistry with Clinical Correlations* in her division for the first time. The product (Wiley PLUS) is an online learning tool for students and a course management tool for professors that provides supplementary materials such as PowerPoint slides, additional study questions, the ability to track grades, the ability to create and assign homework or quizzes, and much more.

Career Path

Darla attended the North Carolina School of Science and Mathematics and very much enjoyed it, but she wanted to try something different in college. She took some business classes during her first two years "and changed majors about a dozen times," she recalls. Then she spent a year away from college and worked in a local business, where she found parts of the business world very appealing. However, she missed the science connection and found herself reading chemistry books in her spare time. She then returned to Appalachian State University and earned a B.S. in chemistry.

Darla realized she "loved research—both the mental aspect (literature, planning syntheses) and the laboratory experience," and went on to study bioorganic chemistry at Duke University. As part of her graduate school experience she coauthored a book chapter; she found the publishing process fascinating.

Toward the end of her graduate school career, Darla received job offers from several pharmaceutical companies, and postdoctoral opportunities at academic institutes. However, she realized she wanted a career that would balance her love of chemistry and her business skills—"in particular, thinking about challenges and developing ideas and solutions to overcome those challenges, presenting ideas and garnering team support, implementing the ideas, and achieving resolution/success for a project." She explains, "Of course, these skills are utilized in a laboratory or academic setting as well, for example, in grant writing, planning syntheses and so on, but I found the challenge of identifying information needs in the chemistry community and looking for solutions to meet those needs very exciting. When the publishing opportunity presented itself, I had to think long and hard about it, because I understood it meant, in all likelihood, leaving the laboratory experience and not returning to that aspect of chemistry—it is, of course, possible, but would be a difficult transition back to the lab."

She interviewed with John Wiley & Sons publishers, and following a few visits to the New York office and a trial run at acquiring content (she recalls, "I conducted market evaluations, wrote up several proposals for new book ideas, then worked with a Wiley editor to pursue authors for these books"), she was offered a position as an associate editor.

Darla had been with the company for nine months when her direct supervisor left to pursue another opportunity. As a result, she was placed in charge of handling all the content and acquisitions for both positions. Darla recalls, "With all this, I was generally working 14 hour days for a year. However, my graduate school lifestyle had prepared me to work under pressure, and I used that experience to help drive me through the challenge. In addition to learning the publishing basics quickly, I learned important life skills during this time, including prioritizing, time management, how to make the best decisions quickly, and how to be a good team leader and manager."

In 2002, Darla had a son and quickly realized the complications inherent in commuting and working long hours. She negotiated with the company to work off-site for a few years (and part-time for the first year of that), so she is now part of the "virtual workplace." This does require coordination, as Darla travels to corporate headquarters for three days each month, and travels for conventions and author meetings another three or four days per month. It also requires dedication and self-motivation, since new opportunities are now her primary driver. While the arrangement is working well, Darla expects to return to the main office in a few years.

As Darla has moved up at Wiley, she has been able to take advantage of many opportunities. Her responsibilities have grown to include participating more in strategic decision making for the division, managing and developing editors and assistants, and overseeing more products and a larger variety of product types. She is also expected to achieve greater financial results for the company. She was able to spend three months working in the Wiley–VCH office in Weinheim, Germany, gaining valuable international experience and exposure.

There are many skills important in Darla's work as an editor. First and foremost, she says, "You must like to talk to people! You need good interpersonal skills—to be able to identify and interact with the right people, to get them to talk candidly with you, and to build long-term relationships.

"Secondly, it helps tremendously to have a background in scientific knowledge, and sound judgment. Scientists are always very enthusiastic and able to sell you on all their ideas—learning to distinguish between these in terms of successful publishing is a challenge. You must learn to be objective—this is a business, and meeting the customer's information needs is what drives the bottom line. You must be willing to do your homework, including research on a specific topic—the market size, current competitors, information needs of that market, workflows, and so on, then make important decisions with the information at hand. Often this requires being very detail-oriented and being able to stay on task until the decision has been reached.

"Of course, excellent writing skills are crucial, and a basic set of business skills and understanding—finance, marketing, sales, Return on Investment

(ROI)—is also important. Blending scientific expertise with business skills and acumen can be difficult, because what is most scientifically interesting may not be the most marketable. Learning to balance the two is critical."

The best part of Darla's job is the interaction with chemists of all career paths, learning about their information needs and devising strategies to meet them. Darla enjoys the growing multi-disciplinary nature and globalization of the entire scientific enterprise. She loves "the excitement, and the challenge." She notes, "I'm doing something different but using and building on my chemistry background."

In the future, she hopes to continue to serve her authors and chemistry customers well by providing good-quality content in the format desired. She also aims to help the company increase market share for chemistry content.

Advice

Darla notes, "There is a shortage of scientific applicants for this field of work, and most publishing companies are recruiting for these types of positions every year. Most applicants come straight from graduate school, because it is difficult to locate someone with both scientific knowledge and publishing experience. As a result, most starting applicants gain publishing knowledge and skills through experience, as they work.

"Change is inevitable—we all evolve! Don't be afraid to learn about or try something new."

Predictions

As for the future, Darla says, "Scientific literature is becoming more complicated and populated, placing importance on understanding the customers' workflow better, so the customer can get to the information desired, when they need it. Staying on top of the most current hot topics requires dedication to following the published literature, which is more often in electronic and informal formats.

"Protecting the rights of both authors and publishers, while enabling access to information, is a continuing challenge. "

ADDITIONAL RESOURCES

American Film Institute Catalyst Workshop (www.afi.com/education/catalyst/default.aspx) is a weekend workshop focused on storytelling and screenwriting

ACS Division of Chemical Information (www.acscinf.org) hosts an electronic mailing list called CHMINF-L (Chemical Information Sources Discussion List)

American Medical Writers Association (www.amwa.org)

American Medical Association (www.ama-assn.org) conducts research and holds conferences on the topic of peer review

Association of American Publishers (www.publishers.org)

Boston College (www.bc.edu), Johns Hopkins (www.jhu.edu), Massachusetts Institute of Technology (web.mit.edu), the University of California–Santa Cruz (ucsc.edu), and the University of Wisconsin all have one- to two-year science-writing programs that are highly recommended

Careers in Science Writing: (www.geocities.com/CapeCanaveral/Hangar/4707/write.html)

Council of Science Editors (www.councilscienceeditors.org)

Editorial Freelancers Association (www.the-efa.org) is a professional organization for freelance editors that posts rates for editorial services at www.the-efa.org/services/jobfees.htm

National Association of Science Writers (www.nasw.org) helps writers improve their craft and encourage conditions that promote good science writing

Society for Scholarly Publishing (www.sspnet.org)

Society for Technical Communication (www.stc.org) and their Scientific Communication (www.stcsig.org/sc) and Environmental, Safety, and Health Communication (www.stcsig.org/esh) special interest groups

Special Libraries Association (www.sla.org) is a nonprofit global organization for innovative information professionals and their strategic partners

TECHWR-L (www.techwr-l.com) is a diverse and active community of technical communicators that has an electronic mailing list and searchable archives

Women in Film (www.wif.org) empowers, promotes, nurtures, and mentors women in the industry through a network of valuable contacts, events, and programs

Writer's Market (www.writersmarket.com) lists publications that contract for articles from freelance writers

Writers Write Writer's Guidelines Directory (www.writerswrite.com/guidelines) is an online database of publications that accept submissions

University of Montana, Bozeman, MT (www.montana.edu) has the only science documentary-film program in the country, a two-year program sponsored by Discovery Science and Sony

2 Chemistry and Information Science

THIS TRULY IS the information age—the sheer volume of resources and information available is overwhelming. Some of the most valuable people in an organization are those who can effectively find, organize, and disseminate the appropriate information and teach others how to efficiently find the information they need. These are the chemical information professionals—science librarians and information scientists. If you have a greater interest in scientific literature than in laboratory work, this may be the career for you. Science librarians are motivated by curiosity and also by a desire to continue learning and to share that learning with others. Science librarians work as part of a team, supporting other scientists. A science background is essential, enabling one to communicate with scientists on their own level. A research background is also helpful in order to understand how information is used for making decisions in chemical research.

Recent technological changes have significantly altered the role of the science librarian, and the physical layouts of many libraries themselves. As specialized libraries merge into single-source resource centers and as more resources become available from offsite locations, the role of the librarian is evolving. Chemical-information professionals today are transforming from the traditional role of gatekeeper and provider of answers to the new role of expert resource, teacher, and enabler. Science librarians today spend much of their time evaluating available resources and teaching others how to conduct searches and much less time researching answers for others. In some cases, especially in large organizations, information specialists are hired specifically to conduct expert searches, which originally was in the purview of the librarian.

* * *

THE modern library is not a musty building full of old books, and librarians don't just put books on shelves and tell people to be quiet. Libraries (or information centers) hold books, magazines, videos, CDs and DVDs, software, and all varieties of electronic resources. They also hold computers that allow access to the Internet and a vast array of remote resources. In addition, many of these resources can be used by patrons in remote locations—they don't even have to come into the library anymore. Many of the resources that patrons now require did not exist a few short years ago, and even the ones that did are constantly changing. The science librarian must keep abreast of currently available and constantly changing information sources. In addition to searching high and low to find a specific nugget of information, librarians are responsible for providing instruction in the use of these tools and teaching patrons to navigate for themselves the flood of available sources.

A number of historical events have brought about this transition. When they were first introduced, most chemical information databases charged by the minute, or by the number of search results (hits). Poor search strategies took longer and returned many more irrelevant hits and so were immediately and quantitatively more expensive. Under these conditions, researchers would ask their questions of librarians, who had special training in using the tools and were able to conduct searches more quickly and cost-effectively.

Over time, several factors have combined to make searching by the non-expert more cost-effective, such as:

- Databases' becoming easier to use and searching more efficiently.
- Networks' allowing access from virtually any location.
- License agreements' moving to a single institutional annual fee for unlimited access.

Additionally, new technology is more powerful—it allows much more complicated searches than were possible with printed indexes only (for example, chemical structure and substructure searches). Since searches are more efficient, preliminary search results can be used to guide further research. Literature searching has become both more powerful and more complex than ever, and ever more knowledge of chemistry is required to use search tools efficiently. For this reason, in the business world a new profession of "information scientist" has developed. Information scientists (or similarly titled positions) provide advanced information-retrieval services; their sole responsibility is shaping queries to retrieve information. And, as a company's information needs expand, the information scientist's role expands beyond traditional information retrieval and analysis into purchasing and licensing information products for the corporation.

Science Librarian

The job title "science librarian" covers a wide variety of functions, from serving as an expert resource for scientists doing original research, to assisting and teaching undergraduate students just learning how to conduct scientific searches. Libraries usually offer public computer and Internet access, so librarians must oversee those resources and often are expected to train patrons in their use. Science librarians must know how to extract the required information from their collection, or where to go if the information is elsewhere. Science librarians have a wide range of duties, and the relative importance of each aspect will depend on the specific position and type of institution.

Librarians, especially those at academic institutions, spend an increasing amount of time teaching information-retrieval skills. In many cases, this teaching is not done in a classroom setting. Self-paced tutorials and distance-based education are becoming ever-more popular because students want to learn at their own pace and on their own time. "Distance learning can be as effective as attending a traditional face-to-face class, but only if there is an expert available to answer questions and provide support," says Bruce Slutsky, a technical reference librarian at the New Jersey Institute of Technology.

Science librarians must continually evaluate new resources and interfaces. For example, there are many competing Web-based interfaces that can be used to access commercial online services, so the librarian must evaluate each and determine which is best suited for their particular patrons by considering factors such as ease of use, cost, depth versus breadth of coverage, and so on. As new resources become available and old ones changes, these factors must be continually reevaluated and the collection continually updated. Access questions are becoming more complicated. Institutions need to allow patrons access to electronic resources from multiple locations but to have to restrict that access to legitimate patrons. Often this involves technical challenges that require significant computer networking expertise or close collaboration with those who have that expertise.

Senior librarians can be responsible for substantial budgets and for selecting not only books and journal subscriptions but electronic databases and resources as well. This necessitates maintaining good relationships with external vendors, and joining consortia or facilitating collaborations to allow access to complementary collections.

Some academic librarian positions come with faculty rank and the associated benefits. However, only about 20% of all universities have a separate chemistry library, and only 50% of all universities have a dedicated science library. These numbers are decreasing as the boundaries between disciplines blur and libraries merge into interdisciplinary resource centers. Today's resource center is as much a collection of services as a physical place. While academic libraries are a small market with perhaps a dozen job openings per month across the country, there are many other places where you can find people doing this type of work. Major companies also have libraries, or technical information centers, where not only books and journals but also corporate intellectual property and knowledge are archived. Many nonprofit organizations, book and journal publishers, scholarly organizations, government offices, museums, science centers, and so on have resource collections to manage. Some may provide resources to the public and use librarians to manage the dissemination of that material. Depending on the nature of the organization and the type of population it serves, science librarians may be expected to deal with inquiries and resources on a broad range of topics.

About two-thirds of all academic librarian positions require a specialized degree, called a master of library science (MLS) degree. "The MLS gives you additional perspective," says Svetla Baykoucheva, the head of the White Memorial Chemistry Library at the University of Maryland, "and those with both science and library–information science degrees are rare." The common background with other librarians makes discussions easier, which is crucial for establishing relationships. The MLS degree is best obtained after some experience in the field and can be earned part-time while working. In fact, there are schools that offer this degree online. In industry, a science background is more important than an MLS degree, especially for the information scientist (as opposed to library administrator) positions, and advanced scientific degrees are often required.

During tough economic times, libraries can be hurt because they are seen as an expense and are not revenue-generating departments. This makes them attractive when costs need to be cut, so librarians must be proactive in ensuring that the value-added nature of their work is appreciated. "Science libraries suffer the same lack of respect as other libraries," says Phil Barnett, an associate professor and science reference librarian at the City College of New York.

The chemistry librarian position will never completely disappear from academic institutions, especially at those universities large enough to require a separate chemistry library. In industry, the trend is toward two separate positions —administrators, who evaluate and acquire information resources, and searchers (information scientists), who make use of the resources (but the pendulum will probably eventually swing back toward a single, combined position). Research chemists themselves do the straightforward and basic searches, but specialized information scientists do the complicated or advanced searches. This is especially true when patent literature is involved—patent searching is a specialty in its own right and is discussed in the following chapter.

Information Scientist or Information Specialist

Research and searching for answers was a large part of the traditional librarian's role. In many places, these functions have expanded into a separate job title, "information scientist." Information scientists work as traditional information searchers, researching answers to specific questions for other scientists. A background in chemistry, or in another science, is essential not only to understand complex databases but to aid in defining and focusing the user's search. Many times the most important part of the process is getting the scientist to articulate what information he or she really needs.

Information scientists not only locate but often interpret, summarize, manage, and organize information, including electronic documents such as Web pages, policies and procedures, newsletters, and so on. They may prepare directories or overviews to let people find information more easily. These directories may need to be organized differently for different groups, so the work requires an understanding of how each type of user will expect to find things. A science background provides the vocabulary, values, and methodology of science and allows the information scientist to communicate with scientists on their own level.

Providing information is a service industry. Information scientists spend a significant amount of time in contact with other people, and the work must be done to their satisfaction. Being people-oriented, approachable, and able to market services to colleagues are all important. An eye for detail and a love of public service are also key attributes. Most training occurs on the job, since the available resources change so fast. In this field, even more so than in many others, keeping up with changing trends and tools is vital for continued employability.

Abstracting, Indexing, and More

Many companies provide value-added information services, such as abstracts of chemical information or specialized databases. By organizing the data and pulling information from multiple sources, they create value-added products that are extremely valuable in their particular niche. These types of companies are called "secondary publishers" because they publish compilations of information that has already been published in another format. (Primary publishers are those who help create the content, such as traditional journals and conference proceedings). Secondary publishers include Chemical Abstracts Service (CAS), BIOSIS for biological information, and MedLine for medical information. Much of the work of these publishers will never be automated, because a computer program cannot tell the difference between, for example, mercury (the element) and Mercury (the Roman messenger of the gods).

Abstractors and indexers used to be employed by these types of companies to prepare and mark information so that both people and automated systems could easily retrieve it. Nowadays, many of these positions have expanded into "document analyst," "scientific information analyst," or a similar title and perform a wide variety of analysis and annotation as described below.

An abstract is a summary of an article's content, which briefly describes the concepts and chemical compounds discussed in the paper, research report, or other document. While most journal submissions require the author to prepare an abstract, in some cases that will have to be revised or rewritten for a new audience. In many cases, metadata is added to the document. Metadata is data about the data—authors, date, source, description, and so on.

Index entries are created, usually from a specific vocabulary list. Consistent use of keywords, which may include adding some not used by the authors, is vital to ensure ready retrieval of all relevant information. The most useful databases are those that are amenable to natural language searches and that provide a controlled vocabulary list (in the form of a dictionary, lexicon, or thesaurus).

Especially for chemical data, both structures and systematic names must be created and stored in a searchable format. Generation of unique names from structures requires considerable skill. Patent specifications, which use generic Markush diagrams to represent entire classes of compounds, require information scientists with the skills to name and index properly. Chemical databases often contain chemical and physical-property data, often including specific numeric values. Biological databases may include the biological properties of the chemical compounds, diseases, and names of plants and animals (again, using a highly controlled vocabulary).

Database Development

In addition to analyzing data, information scientists are trying to manage the flood of information by organizing it into databases, or collections of similar information in easily searchable formats. Resources such as the Protein Data Bank (PDB) and the Cambridge Crystallographic Data Centre (CCDC) provide not only vast repositories of experimentally determined data but also software tools to mine information from the data stored there. The CAS Chemical Registry System includes millions of chemical compounds, identified by systematic names, trivial names, chemical structure, and much more, along with software tools to facilitate searching and analysis of the data.

However, databases of chemical information can be more complicated than most, because not only physical data and text but also chemical-structure information can be included. Chemical data may need to be indexed by two-dimensional and three-dimensional descriptors, International Union of Pure and Applied Chemistry systematic names, Chemical Abstract Index names, common names, trade names, and much more.

Many times, databases are created to facilitate particular kinds of searches. For example, a database will be created for similarity searching, which locates compounds similar to an existing structure. However, how is similarity between chemical structures defined? Three-dimensional shapes are hard to quantify, and comparing all possible orientations of all possible compounds is beyond the computational capacity of most companies. To further complicate the matter, searches may be required for similar or complementary sizes and shapes, or for similar structural or molecular property characteristics. Storing and marking the data in such a way as to make this possible is a continuing challenge.

Profiles

Patricia Kirkwood

Engineering and Mathematics Librarian
University of Arkansas

BS, Chemistry, Pacific Lutheran University, Tacoma, Washington, 1980
MS, Library and Information Science, University of Illinois, Champaign–Urbana, 1984

Current Position

Patricia Kirkwood is currently the engineering and mathematics librarian at the University of Arkansas, a large state flagship university with an active scientific research community. Her job responsibilities fall into three major categories. First is collection development, which includes ordering books, arranging vendor visits, reviewing materials (both electronically and on paper), and using spreadsheets to track costs. She spends about 20 hours per week on these tasks. Patricia is also a faculty member and spends time each week instructing students in how to conduct literature searches, through both formal class instruction and individual sessions; demonstrating information tools to faculty members; preparing handouts for classes; and working with students on projects and with researchers on research questions. Finally, she spends five to ten hours per week working at the reference desk or tracking down information for patrons and answering questions via e-mail and through a live-chat reference application. She has a second office in the engineering building near her primary users, which allows her to interact informally with the faculty and their research groups.

Patricia says, "One of the best things about being a science-based librarian is that I get to work with technical people. Scientists—which include engineers—are interesting people. There is a wide variety of types, but they are all very interested in their work. They love to share what they are doing, if you take the time to ask, and their information needs provide a variety of opportunities for me to learn about the science that is shaping our world. For example, it was a real treat when a Nobel laureate in chemistry took the time to invite my library's staff to an artistic and literary display with which he was involved. This ended up in a two-hour personal tour of some marvelous art associated with science.

Recently, a well-known engineer joined the students at a pizza party as we reviewed a new engineering database.

"Working with the faculty and students is great fun and a new challenge every week, and working with fellow librarians who know their fields and share my enthusiasm for teaching and providing information is a joyful experience. The people I work with know the value of information and are generally supportive. Even the biggest thorn, and there are some difficult professors and postdoctoral scientists, can become a wonderful supporter of the library over time.

"There are some challenges involved in science librarianship. Knowing that the price of information is always going to outpace the funding available is an annoyance, but I try to accept this as a challenge. I figure out how to get the best resources for my institution with the funding we have. Then I make sure the students and faculty are aware of the wealth of resources available to them.

"Technology is another issue that has both a best and worst side. It is constantly changing. There are days when I would like all the vendors, software providers, and database managers to just *stop*! No changes for six months. No new interfaces. No new products. But then I catch my breath or hear of something really unique that can solve a problem I've seen and I'm off to the races again."

Career Path

Patricia started her career with an undergraduate degree in chemistry and then obtained an industrial position as a laboratory-bench chemist. After about three years, she found herself bored but didn't want to go to graduate school for a PhD. She started investigating her career options and realized that she really liked the information research aspects of her job. She decided on librarianship and found that with only one more year of college she could get the credentials needed to make it a career. She went back to school and obtained an MS degree in library and information science while working in various libraries on campus. She notes, "This opened up a world of options—I went from having a good-paying job with little mobility to a fantastic career with lots of options."

After obtaining her degree, Patricia went to work as a reference librarian at Bell Communication Research, where she remained for two years. While there, she realized she missed teaching. So she decided to transition from industry to an academic library.

Patricia was a new mother at the time, and the increased flexibility and more rural setting of an academic position appealed to her. After looking around, she obtained a position at Emory University in Georgia. She says, "If I'd stayed in industry, my focus would have been more on training, immediate service, and, perhaps, patent searching. As an academic librarian, I was able to run a small departmental library, teach, and develop a good collection for researchers. I had a lot more freedom to develop professionally in the areas I was interested in, which were teaching and long-term collection development." After a time, she was promoted to science librarian, which meant she supervised two other librarians.

After a few years, Patricia moved to a larger school (Cornell University), where she ran a medium-sized library. Again, she was able to develop instruction programs and work with researchers to develop a robust chemistry collection. Patricia recalls, "I really learned how to search chemical abstracts while in this position. And the chemistry department paid for some of my volunteer activities with the American Chemical Society—a nice benefit!"

She says, "Then life happened. Between health and family issues, and management changes at the university, I decided it was time for a change." She discovered an opening for a general-science librarian at her alma mater, Pacific Lutheran University in Tacoma, Washington, and applied. Although it is unusual for a specialized librarian at a large school to become a generalist at a smaller school, Patricia welcomed the opportunity to give back to the institution that had given her so much as an undergraduate. At Pacific Lutheran University, she was able to spend a considerable amount of time developing an undergraduate instruction program. She adds, "I was able broaden my subject and collection-development skills to include nursing, engineering, biology, and—of all things—music. These skills and areas were not what I had planned to develop, but they gave me a considerable amount of flexibility in terms of career progression."

This extended expertise came in handy when she decided to move again. Patricia wanted to teach more students (classes of 50 to 100, instead of 5 or 10), have the opportunity to do research, and be able to develop a collection. Patricia is now an engineering and mathematics librarian at the University of Arkansas. She is "back into working at a research library, but in a different field—engineering." She says, "The opportunities for interaction with faculty and students, and for collection building are amazing—and learning about engineering in depth is intellectually challenging. Almost a year after the move, I feel that both personally and professionally I made the correct decision."

As for the future, Patricia has many possibilities. Being relatively new in a position that was vacant for some time allows her to shape the position the way she wants. She anticipates some opportunities for grant writing, to obtain funding that will allow her to modify the current library-instruction programs in engineering at the University of Arkansas. This could lead to some exciting research into new methods of teaching. In the next few years, her family obligations will lessen, and she may use that freedom to travel more within her current work environment. Her administration is very supportive of professional travel and advancement, so this is a real possibility. If she chooses to move again, there are many possible career paths available.

As for what she would have done differently, Patricia says, "I think if I had started my career at an undergraduate institution I would have had a smoother transition to the big-league schools. I should have started small and worked my way up. With a degree in chemistry and an MLS, there are many opportunities available in the academic world. Unfortunately, they often come with administrative and management responsibilities that need skill sets outside the normal training parameters." Knowing that ahead of time, and preparing by learning about budgets, personnel management, and politics would have been helpful. However, learning on the job has its own rewards, and overall she's pleased with how her career has progressed so far.

Advice

Patricia says, "I think of the science librarian as a research partner who does not have to be in the lab. I get to work with many interesting, intelligent, and resourceful people, which provides me with many opportunities to grow and be challenged.

"The most important skill I use is the ability to listen and hear what the user is really looking for. Patience, persistence, and the ability to concentrate in a busy environment also come in quite handy. Facility with a computer is very important—the more computer skills you have, the better off you will be. Intellectual curiosity, especially about technical topics, keeps the work interesting, and the ability to know a little bit about a lot of things makes it easier to find the right starting point. Finally, a willingness to be flexible in an environment that is constantly changing allows you to stay sane."

Many aspects of a chemistry background come in handy in an information-science career. Patricia notes, "Of course, basic scientific and chemical knowledge are essential—you must know enough to ask intelligent questions and be aware of the terminology and the organization of science." She recommends "considering a master's degree in your specific subject, especially if you are interested in the more-biological areas. The in-depth knowledge of biological systems may lead to very exciting information-profession opportunities in the future. However, you must also learn how to write and communicate with nonscientists. The science librarian is often 'sitting on the fence,' communicating between the science community and the library community. Sometimes the values of these two groups are not in step, and translation is needed.

"Analytical reasoning and an understanding of the research process are also important. The ability to take a look at a problem, define the areas and issues, and develop a solution comes directly from chemical-laboratory work. You must also have persistence—keep questioning and investigating until you know you have the correct answer."

Predictions

"As tools like Google become more robust and better able to serve the casual user," Patricia observes, "librarians need to be even more involved with the creation and organization of information. Online tools don't work unless the organization is present. Notice the move to Google Scholar and Sircus. In addition, librarians will need to work with vendors to develop products that are worth the price and to educate users as to when 'free' means 'poor-quality' and 'fee' means 'saving time and better quality.' This is one of the biggest growth areas for librarians in an academic setting. How do we teach information literacy and competency in an age when the computer and the Internet seem to answer all the questions?"

David A. Breiner

Manager, Technical Information Center
Cytec Industries, Inc.

BS, Biology, Fairfield University, Connecticut, 1988
MS, Chemistry, Sacred Heart University, Fairfield, Connecticut, 1999

Current Position

David is the manager of the Technical Information Center (TIC) at Cytec Industries. More than just a library, the Technical Information Center is an institutional memory, a repository for intellectual property belonging to the company. The TIC began as part of a knowledge-management initiative in 2003 and was designed to be a center for idea exchange and innovation in addition to employee learning. The TIC archives laboratory notebooks, research reports, monthly reports, and so on—all the intellectual property that is valuable to the company. "Preserving the rich corporate history is essential. As people retire, the company needs to keep their knowledge and expertise. It must be documented properly, so you need to have a process to make sure that it gets captured. This is what we are working on," says David.

In addition, managing intellectual resources from a central system allows easier collaboration. This central clearinghouse for technical information not only helps build an institutional memory but also allows employees to see what's going on in different departments. One way they encourage collaborations is by sponsoring Friday afternoon poster sessions, with scientists from different departments making presentations. David explains, "Often, different departments don't know what each other is doing, so we try to promote interaction between departments on a technical level. We try to connect people to people and promote information sharing."

After spending his first year on the job setting up the infrastructure, David now spends most of his time convincing research groups to take advantage of his services and allow their documents to be managed as part of the TIC. He talks to scientists, managers, and directors daily, both formally and informally. He must convince them that they should think about other aspects of the data or think about intellectual property in a new way. The first step was getting everyone to switch from paper to electronic journals, in order to support multiple sites without having to purchase multiple physical copies. This was a huge paradigm shift but was very exciting. David notes, "We are a global Research & Development organization, so having everything available electronically just makes the most sense in today's digital age. Switching to electronic journals from publishers such as the American Chemical Society and Elsevier was a great way to start."

Within the TIC, some documents are public. For example, public information, such as patents, presentations at outside conferences, marketing materials, and so on, is available to all employees. Reports are all completed electronically and stored. As part of the approval process for a conference presentation, the employee must submit a copy of the presentation, which is then archived. Electronic copies of notebooks will be soon be added to the archives, and they will begin exploring electronic laboratory notebooks in the next few years.

However, there are security concerns—not every employee can, or should, be able to view every document. Research and development leaders decide who can see what information and how permissions are set on specific files and folders. The default setting is for each employee to be allowed to access only documents within their own department. Those higher on the corporate ladder have more access, but the level of access depends on both the document and the individual requesting access. This central document-management system has the strong support of upper management, as evidenced by the fact that David originally reported directly to the vice president of research and development. This allowed him to make quick decisions and do things faster, without having to fight levels of bureaucracy.

The TIC also serves as a traditional corporate library, so David has many of the same duties as a traditional librarian—managing the budget, talking to vendors of various information services, working out subscription levels and costs, acquiring books and electronic resources, and so on. He manages the internal library Web site, which includes electronic journals, patents, market research, online books, and other external information, and links to an internal document-management system. His responsibilities include not only maintaining and updating the site but also providing support and training to users on how to use the various resources.

David does some literature searching when his associate needs help, and the TIC outsources some searches as well. He says, "This is often easier and faster than doing it ourselves, especially when the searches are extremely complex and go beyond our level of expertise. We have enabled the researchers to do basic searches themselves, and this allows us to spend more time on the difficult searches, such as prior art searches for patentability and infringement studies. It's sometimes more cost-effective for us to contract some searches out, especially if we are working on several requests at the same time."

Career Path

During his undergraduate career, David was a premed biology major, with minors in chemistry and communications. Upon graduation, he started medical school, in part because he couldn't see himself working at "a job." During the second year of medical school, the workload became too grueling. David realized that medicine was a lifelong, 24–hour commitment that he was not willing to make. He left after the second year and did not want to go back to school.

However, he soon learned that no one wanted to hire a medical school dropout. He ended up working as a waiter for more than a year. He didn't really want to work in a lab but didn't know what he was qualified for or what he really wanted to do. He interviewed for pharmaceutical sales representative positions, but competition was fierce. David recalls, "I scoured the Hartford paper every week, but 100 people would apply for one job." The absolutely lowest point was an interview for a position selling photocopiers. He was offered the position but couldn't bring himself to take it.

Finally, David saw an ad for a position at Chemical Abstracts Service (CAS). He applied and was interviewed in Columbus, Ohio. He was offered a position as the Northeastern account representative doing technical sales. The position involved calling on established and new customers and explaining the available packages, as well as providing introductory training. David accepted the position and traveled extensively over the next two years. He serviced accounts throughout New England and New York, trained chemists and information professionals in STN Online searching skills, and developed new business accounts for CAS, including many in the biotechnology sector.

After about two years, a manager from one of his accounts called, said that one of their searchers was leaving, and asked if David knew anyone who might be interested in filling the position. They wanted to hire an entry-level person and grow them into the role of information scientist. David realized he was tired of traveling constantly, so he leveraged his expertise in training chemists in CAS/STN online searching and was offered the position. He moved to Bayer Healthcare and continued to support their CAS/STN online searching and learned a wide variety of other tools for chemical, biological, and patent searching. During this time he also went back to school and earned a master's degree in chemistry, got married, and had his first child. While he enjoyed the position, he realized he never wanted his boss's job and didn't have the experience required to do it anyway. Since moving up at his present company was not an attractive option, he eventually began to look around for other opportunities.

In 1998 Bristol-Myers Squibb (another of his former customers) called, said their senior searcher was leaving, and asked if David knew anyone who might be interested. David was, and he took the new position. The new job was at a higher level than the position he had and focused more on patents, an area in which he was interested in expanding his skills. He was able to expand his skills in other areas, too, such as administering and supporting CAS SciFinder across five U.S. research sites and designing an internal Web site for patent information. After a couple years, he moved from searching into the knowledge-management group. His role was to design and implement Web sites for more than 100 global project teams. While he enjoyed the work, the internal politics were unpleasant. Over time, he noticed that the company was moving everyone in his department into project-management roles, which was really not what he wanted to do.

In 2003 David got another phone call. This one came not from a former customer but from a recruiter. David firmly believes that you should "never refuse an interview—always go to find out what they have to say. You either learn that you have it good, or find out about something better. The interview can be its own good experience." David went on the interview and found out about something better. Cytec Industries, a specialty chemicals and materials company, was looking for someone to turn their library into a technical-information center. More than just a technical resource, it would collect and manage all intellectual-property knowledge within the company. They needed someone to start at the beginning and build the infrastructure. David was ready for a challenge and accepted the job. He says, "Moving from a department of 70 people to a department of 2—myself and another person to be hired later—was a big change. I was able to become a manager of people, not projects—and the salary increase didn't hurt either."

It has been a challenging position. He explains, "I was given a clean slate, and I got to build my own infrastructure. I don't have a huge staff, so I have to learn how to do many things myself, but I get to learn lots of new skills in the process. There's always something interesting going on and I have to tear myself away at 5 P.M. After two years of work, we have a nice story to tell of what we've been able to do, growing our resources and our reach while still coming in under budget. We have proven the value of the TIC to the company."

There are many skills David finds useful in this position. He notes, "Technical skills—using documentation-management software, being able to conduct efficient Web searches, and so on—are good, but people skills are crucial. You have to talk to people and communicate well with them. Being able to adapt to change quickly also helps."

If he had things to do over, David would have "been more serious in college," thinking about what he wanted to do with his life. He says, "I was a serious student, but I didn't think long-term. I just figured it would work itself out after I got out of school. However, over the course of my career I took the opportunities that felt right at the time. I'm happy now, so I must have done something right."

Advice

David advises, "Learn new skills at each job. Know when to leave—when you get comfortable, you're not being challenged and not learning new things. Here I'm slowly starting to serve other divisions, but that involves selling the idea to them, so I'm learning how to do that. Overall, I'd say find the things that are exciting to you, the things you are passionate about. Then find a way to make them your career."

Predictions

As for the future, David says, "I believe that the need for information specialists will become increasing important as we move more into a 'digital' world. There will always be a need for those who can make sense out of information and help others turn it into knowledge."

W. Val Metanomski

Senior Scientific Information Specialist in Authority
Database Operations Department, Chemical Abstracts Service (CAS)

BSc, Chemical Engineering, University of London, England, 1952
MSc, Chemical Engineering, University of Toronto, Canada, 1960
PhD, Chemical Engineering, University of Toronto, Canada, 1964

Current Position

Val Metanomski is a senior scientific information specialist with Chemical Abstracts Service (CAS). His current responsibility is to generate and edit Chemical Abstracts Index names for polymers selected by other specialists from the primary literature (journal articles and patents) for inclusion in the CAS database (Registry File, CA File, etc.).

On a typical day, Val will examine several queues of chemical structures, sorted and accessible by the CAS registry numbers, at various stages in the naming process. Chemical substances (compounds) are selected for indexing by document analysts who read full original journal papers, conference proceedings, and patents. Compounds that are new, or have new information reported about them, are selected for inclusion in the databases.

Once a compound has been selected, Val's job is to generate the CA Index name, according to the CAS nomenclature rules. First, the compound's structure is recorded and compared to existing ones, to determine if it has already been indexed. If no match is found, a name is initially generated by either an editor or a name-generation algorithm. Contrary to popular belief, CA Index names are types of IUPAC names yet are unique (IUPAC allows alternatives). The names are almost totally systematic (IUPAC allows many trivial names). They are formatted (inverted) to suit the requirements of an alphabetically arranged index. The generated name is then checked by a second editor to ensure accuracy. For example, an editor will check generated names against connection tables. The entire process, from selecting a compound to releasing the verified name, can take a month.

Val explains, "Polymer—macromolecular—nomenclature is quite challenging, and rules have to be continually updated as new types of polymers are synthesized. Polymers are actually given two types of names, source-based names expressed in terms of monomers—starting materials, and structure-based names expressed in terms of structural—constitutional—repeating units, if known."

Career Path

While Val did well in chemical engineering undergraduate studies, he probably would not have chosen this career if World War II had not interrupted his high-school studies. He obtained a high-school diploma while on a short leave from the military, but college studies had to wait until the war was over and he was discharged from the army. At that time, available grants were given only to those pursuing careers in practical sciences, chemistry, engineering, and so on, so Val chose chemistry for his undergraduate major.

Val became interested in chemical literature, information retrieval, and related problems as early as the fourth year of his undergraduate studies in England. He recalls, "At that time there was a requirement to produce a thesis to earn a bachelor of science degree. This required considerable literature search and understanding of prior art. I really got hooked to it and visited the prestigious Patent Office in London a number of times."

Having received a BS in chemical engineering, Val moved to Canada and worked for a water-treatment company that specialized in protecting corrosion-prone industrial equipment, especially boilers, cooling towers, and pipes. For two years, he worked in their laboratory in Toronto, analyzing samples of water, industrial scales, cleaners, protective tapes, and so on. For another four years, he worked for the same company as a service engineer, mainly in southern Ontario, traveling to the customers' sites to analyze water on the spot, inspect boilers, and make recommendations to prevent corrosion.

After six years in industry, Val realized his current position did not provide much satisfaction or opportunity for advancement. He wanted to get ahead and felt that a doctorate would help. He decided to go back to school and continue in the same field toward a PhD. He recalls, "At the time, I had no idea that when I finished my graduate studies I would proceed to an information-science career. Originally, I hoped for an academic career. But, when I got the PhD I found I was too old to compete with brilliant and eager young assistant professors.

"Quite early in my chemistry training I became very familiar with the best-known secondary chemical publication, Chemical Abstracts (CA). While in graduate school, I responded to an advertisement from Chemical Abstracts Service (CAS), which was seeking volunteer abstractors. For about six years, I prepared abstracts at home for CAS and thus became well acquainted with the organization."

Upon receiving his doctorate, Val again responded to an advertisement from CAS, this time for a full-time position located in Columbus, Ohio. While this appeared strange to his colleagues, including his PhD advisor, Val already had experience in laboratory, industry, and university positions and had concluded that he would most enjoy a desk job related to chemical information. Val says, "The bottom line is that I never felt comfortable in the laboratory or in the plant, and my disposition was basically to have a desk job. Yet, the chemistry training and practice was essential to my success in the chemical information field."

Val interviewed and was offered the position and has been working for Chemical Abstracts ever since. During this time, his career path has paralleled that of the organization, which went from the printed-product environment to the modern online and Internet environment, and he has adapted to and embraced the changes as they came along. His first responsibilities were abstracting and indexing the polymer literature, because of his graduate education in that field. He was also responsible for creating and applying CA index names for polymers, and eventually this became his special area of expertise.

As time went on, he became involved with a variety of projects aimed at converting the traditional manual operations into highly automated processes. For example, Val assisted editors in developing vocabulary-control systems for both abstracts and index entries. Ensuring that a consistent vocabulary was used throughout all publications simplified the automation and electronic processing of documents.

Val has also been responsible for managing an editorial planning and development department. He took an active part in the transition from the printed-product environment to the modern online and Internet environment of chemical information.

Of his roles in information science, Val says, "It was all immeasurably helped by the acquired basic knowledge of chemistry and engineering, and laboratory and industrial experience. You cannot organize the databases of chemical information and their delivery routes to the 'bench chemists' or 'practicing engineers,' if at one time or another you were not at the receiving end of chemical information."

Throughout his career, Val has also been active in various capacities in the ACS Division of Chemical Information and in the International Union of Pure and Applied Chemistry (IUPAC), especially in the development and application of polymer nomenclature.

One of the biggest challenges in this field is keeping abreast of the tremendous developments in molecular modeling, simulation, data mining, and so on. Another challenge is maintaining familiarity with countless resources on the Internet but with a critical eye to their authenticity, genuineness, and reliability.

At this point, Val expects no changes in his career. He has found a niche in applying his expertise to generating and editing polymer names, which will allow him to work quietly and efficiently long after the formal retirement age. He still interacts with other chemical-information professionals, bench chemists, engineers, computer scientists, and their management and associations.

Val's 40-plus-year career with one company was typical of his generation, but today's environment is much more mobile and more technologically literate. If he had to do it over again, Val would have started studying computer hardware and software much earlier. He admits, "That of course does not apply to today's younger generation, since even elementary school students are quite adept in using terminals to play games, retrieve things of interest, send and receive e-mail, and so on." However, keeping up with technology, no matter how it changes, and being open to incorporating new tools into the workplace are as important as ever.

Advice

In order for one to have a successful career in chemical information, Val feels that "knowledge of the tools available for chemical-information retrieval, and knowledge of chemistry to assign potential-value ratings to the information retrieved" are crucial.

For careers in "chemical information," or "cheminformatics," Val recommends first getting a basic degree in chemistry (attending a three-year college program leading to a B.Sc. would be fine), then a master's degree at an information-science or librarianship school.

Especially people who come to this field from graduate schools (cheminformatics, computer science, librarianship) can expect salaries comparable to those of people who stay in the laboratory.

Val notes, "Membership and participation in the activities of professional organizations are essential. Attending the meetings, symposia, and seminars provides the best insight into the most-recent developments and applications in the field. Making personal contacts with peers, and participating in discussions and exchange of experiences and ideas, face-to-face, in the meeting rooms and the corridors and at the social events, contribute to a better knowledge of the profession and its practitioners."

Predictions

Val is optimistic about job prospects in the field, explaining, "Although most of today's chemists and chemical engineers have at their disposal information-retrieval tools that used to be available only to information professionals, there will still be a need for specialists, within organizations or outside them, whose knowledge and experience will cover *all* the available tools and who will also be able to critically evaluate the available information.

"While the availability of information—databases, Internet, and so on—has increased a thousandfold or more, there is now more unreliable, useless, and downright wrong information at everybody's fingertips. Hence, future professional will have to be able to add value, or old-fashioned relevance, tailored to the specific client's needs."

Bonnie Lawlor

Executive Director
NFAIS (formerly the National Federation of Abstracting and Information Services)

BS, Chemistry, Chestnut Hill College, Philadelphia, Pennsylvania, 1966
MS, Chemistry, St. Joseph's University, Philadelphia, Pennsylvania, 1976
MBA, Wharton School, University of Pennsylvania, 1989

Current Position

Bonnie Lawlor is the executive director of NFAIS, a member organization of information providers that serves the needs of all those involved in any aspect of information creation, distribution, and retrieval. NFAIS membership comprises international scholarly associations, public and private companies, libraries, major corporations, and government agencies. The main activities of these groups are primary and secondary publishing, host systems, technology innovation, data creation, and information distribution. NFAIS facilitates the exchange of information among its members; sponsors topical conferences, seminars, and educational courses; publishes newsletters, current awareness alerts, and books and reports; and develops codes of practice, guiding principles, and white papers on information policy and new technologies.

As the executive director, Bonnie is responsible for recruiting and maintaining member organizations, developing new programs and services, organizing the annual conferences, administering all office functions, and ensuring the overall financial health of the organization. She also maintains an awareness and understanding of trends within the information industry (including content, technology, and policy) to keep members up to date. Which task is emphasized depends on which major NFAIS event is imminent. A typical week will include at least some aspects of developing programs, reviewing current industry events, communicating with members, and dealing with business issues.

Program development involves determining the theme for the annual conference and those for the three or four one-day information sessions that are held throughout the year. Topics may include information policy, copyrights, new technology, automated indexing, private sector and public sector concerns, best practices, Web 2.0, and so on. Once a theme is set, Bonnie and her staff must develop the program itself, identify and invite speakers, arrange for a venue and caterer, set prices, initiate promotion, create handouts, and so on.

Each day, the current awareness and news feeds are reviewed for issues of interest to members, such as copyright legislation, competitive information (What are Google, Yahoo!, Microsoft and others doing today? What new technology has been announced? Are there any new mergers and acquisitions or any new products and services?). News of critical importance is then sent to members via an e-mail listserv, and less-critical information is posted to the Web site.

Each day Bonnie works on the monthly industry analyses on issues such as open access, emerging trends, and so on, and on the bimonthly newsletters for both members and nonmembers.

On any given day, issues related to board meetings, invoicing and bill payment, development and review of financial statements, meeting with auditors and accountants, equipment and supply needs, and so on, are addressed. In many ways, Bonnie is like the owner of a small business, for whom every aspect requires her immediate attention.

Since NFAIS is primarily a resource for members, Bonnie must also field member questions every day, resulting in the requisite research and all that it entails. Bonnie makes personal contact with all member organizations monthly, to ensure that their industry concerns are being addressed. New-member recruitment is ongoing, and she contacts potential new members, sends them information, and so on.

Career Path

Upon graduation from college, Bonnie started working toward a PhD in chemistry at the University of Pennsylvania. However, after completing her coursework, she had to leave the program for personal reasons, so she began searching for a position as a chemist in the Philadelphia area. Most lab positions at the BS level at that time were basically for test-tube washers, so she decided to cast a wider net. She saw an advertisement in the *Evening Bulletin* for a chemical indexer at the Institute for Scientific Information (ISI, now Thomson Scientific) and applied for the position; they hired her. She recalls, "I must admit I had no idea what it would entail, or where it would lead me. And indeed, that position started me on an interesting, challenging, and enjoyable journey into publishing."

Bonnie spent her initial years at ISI as an indexer—reading the current chemistry journals with the purpose of summarizing, according to the indexing rules, all articles that reported new compounds and new reactions. She notes, "I thoroughly enjoyed the work from the very beginning. As an indexer, I was exposed to all of chemistry—I was not limited to a narrow specialty field. I could enjoy the intellectual challenge of science and not—for me—the drudgery of lab work that existed at that time for those without a doctorate. The longer I worked in publishing the more I loved it." She had intended to retire when her husband finished his degree and started working but found she enjoyed the work too much to quit.

In fact, the exposure broadened her scope of chemical knowledge so much that she decided to complete her MS in chemistry at St. Joseph's College (now St. Joseph's University). While there, she was able to convince the chemistry department chair to create a program for chemistry students who were focusing on non-laboratory careers.

Within a few years, she was promoted to supervisor of the indexing group at ISI, then to manager, and shortly thereafter to director of chemical information services. During those early years, ISI's founder, Dr. Eugene Garfield, ran the company. Bonnie remembers, "If you did a good job, he offered you even greater challenges, and this provided me with several opportunities to obtain approval to create new products, including the Current Chemical Reactions database."

Also, each promotion offered the opportunity to learn and build new skills that could be used in a variety of positions—and companies. Bonnie says, "Probably the best opportunity was when the company was going through a management transition. I was able to leverage their needs and convince them to support me through the Wharton Executive MBA program. That degree provided me with the training I needed to do my job, and management credibility to the 'outside world' in the event that I would I seek employment elsewhere."

Fifteen years after joining ISI, Bonnie was promoted to vice president of operations. Over the next seven years, she learned about the publishing and editorial processes beyond ISI's chemistry-related products, took on greater responsibilities, and was promoted to executive vice president of database publishing.

In this role, her publishing expertise was expanded to include managing intellectual property and copyright issues, relations with primary publishers, and all content-licensing agreements with customers and suppliers. She was responsible for transitioning all the print products into electronic editions and was able to participate in corporate acquisitions to expand the company's portfolio of both content and technology.

Bonnie left ISI about two years after it was acquired by Thomson, and shortly thereafter she moved to Michigan to join UMI (now ProQuest), an online information service. She served as the general manager of their Library Division, and while there spearheaded the growth of their digital products and services.

Bonnie left UMI and returned to Philadelphia but did not immediately take on a new position, primarily because of family health issues. She recalls, "I was fortunate enough to be in a position that allowed me take on the role of caretaker. As things improved, I realized that it was best that I remain in the Philadelphia area—at least for a few years—so that I could continue in that role. But I wanted to utilize my skills, and when NFAIS needed a new executive director, the timing was right. I had served the organization as a volunteer for many years—as president and on committees, and it was a good opportunity." Bonnie now brings all of her experience to bear in her current position as NFAIS executive director.

Her current position "requires a thorough understanding of the Information Community: the players, the issues, the challenges, trends, and so on, both here and abroad." She explains, "It requires good communication skills, both written and oral, since public speaking, fundraising, member outreach, and article writing are all part of the job. Editing and proofreading skills are necessary for the publication side of the business. Management experience is required, along with strong financial skills, as the executive director is essentially the financial officer. Normal business software capabilities such as Excel, Word, and PowerPoint are required, along with computer skills and Internet search capabilities. An ability to plan and attention to detail are also necessary since we simultaneously juggle the organization of multiple meetings and publications."

Bonnie interacts with a diverse cast of people who are involved with just about every aspect of information creation and dissemination within the public and private sectors worldwide and at all management levels: company officers and executives, directors, and managers, on down. In the course of a week, she may speak with members of the press, Internet bloggers, library-school faculty and students, primary and secondary publishers, host systems, technology providers, indexers and abstractors (and aspiring indexers and abstractors), and all the usual contacts for small businesses—banks, sales personnel, and so on.

Bonnie enjoys many aspects of her career. She says, "I love the variety of issues that have faced publishing over the years—particularly with regard to intellectual property and the transitioning from a print to an electronic environment. Content ownership, pricing, the issue of photocopying, then downloading, and now the challenges of content distribution via the Internet—all have made for an intellectually challenging career."

As for challenges, there are always new ones. "For example, as a result of all the mergers and acquisitions in the publishing community," Bonnie says, "there is a declining number of potential member organizations. Many of the organizations that were pioneers in the communication or publishing of scientific information—ISI, Derwent, BIOSIS, Engineering Information, Dialog, BRS, and so on—are now either part of a much larger organization, or gone. Furthermore, many organizations simply do not support the level of professional volunteer activities among their employees that was commonplace more than a decade ago, so we're having a harder time finding not only member organizations but also qualified volunteers."

Looking back, Bonnie realizes, "Throughout my publishing career, my science background was often brought into play when developing the editorial content of new products and services. But invaluable to me has been the scientific thought process and deductive reasoning that has allowed me to work through problem solving—whether it be in the development of new pricing models or 'what-if' scenarios for major business decisions."

If she had a chance to do things over again, there are two things that she would have done differently. She says, "First, I would have completed my PhD, as a doctorate in chemistry commands more respect in the scientific field. And second, after completing my MBA, I would have gone to law school in order to focus on the legal side of publishing—intellectual property, contracts, licensing, and so on."

Advice

Overall, Bonnie believes that "the diversity of the publishing field offers many career opportunities for all types of personalities, work preferences, and family situations. One can choose a technical or management track. One can find work suited to the outgoing personality (customer service, sales, etc.) or the introverted one (writing, indexing, programming, etc.)." She notes, "I have had the opportunity to travel the world representing the companies for which I worked, whether it be for conferences and events such as the Frankfurt Book Fair or to meet with peers in other organizations. And with experience and proven ability, freelance, flexible, work-at-home opportunities also arise.

"Depending upon what aspect of publishing is chosen as a career entry point, salaries may be slightly higher—programming—or slightly lower—indexing or abstracting—than laboratory careers. Over the long haul, I suspect they even out at all the various levels. An executive in a publishing firm will make more than a writer, but then an executive in a chemical firm makes more than a researcher.

"A chemistry degree opens many doors—and they are not all to laboratories or the classroom. The path I chose does not necessarily lead to where I am now. Several indexers that I worked with early on are still indexing—they love it. Others have moved on into other aspects of the publishing process such as programming; science writing—for the press, for journals, and freelance; technical writing for product manuals; translations of scientific material; marketing; sales; product training; product development; customer service, and so on. Some have transitioned into law to become experts in intellectual property, copyright, contracts and licensing, mergers and acquisitions, and so on. And others have remained in management at a variety of levels.

"If you love the practice of chemistry, do not consider publishing. If you love the intellectual aspect of science and want to apply that knowledge to a variety of career possibilities, then publishing is one avenue to pursue. One does not have to begin as an indexer as I did. One can begin in sales, or one can take computer-science coursework and get involved in programming or product development. One can get management experience on the job or obtain an MBA—which I strongly recommend doing, ideally part time while getting on-the-job experience at the same time. If a person has really strong and creative writing skills, that may be the entry point—look where it took Madeleine Jacobs Executive Director and CEO of the American Chemical Society. Be realistic about your strengths and weaknesses and about your preferences for a work environment. Knowing those, you can begin to identify how you can best utilize your scientific background in the information field."

Predictions

Looking ahead, Bonnie says, "The creation and dissemination of scientific information will always be with us. How it is packaged, disseminated, and accessed will evolve as various factors shape the flow of scholarly communication. But products and services built around scientific information are here for the long term.

"That said, those involved in publishing today face some serious issues. Traditional publishers, whether they be nonprofit or for-profit, are challenged with developing new business models that are in alignment with the expectations of today's information seekers—models that will not destroy existing revenue streams. The Internet has lowered the barrier to entry to publishing. Like it or not, traditional publishers must create new products and services that compete effectively with Google, Yahoo!, Microsoft, and other new entrants who seek to dominate the information landscape. Open-access publishing continues to shape user expectations with regard to information access and cost. The convergence of four factors continues to shape the information industry: technology, the economics of information creation and dissemination, intellectual property policy, and an increasingly information- and computer-literate society."

ADDITIONAL RESOURCES

ACS Division of Chemical Information (www.acscinf.org) is a source for maintaining professional competency in information resources, information technology, and information policy

American Society for Information Science and Technology (ASIST) (www.asis.org) is a society for information professionals leading the search for new and better theories, techniques, and technologies to improve access to information

American Society of Indexers (www.asindexing.org) is a professional organization devoted to the advancement of indexing, abstracting, and database building

Association of American Publishers (www.publishers.org) is a trade association of the book-publishing industry

Association of Independent Information Professionals (www.aiip.org) provides a wealth of resources for those who want to know more about the information industry and independent information professionals

Careers in Chemical Information (mail.sci.ccny.cuny.edu/~phibarn/careers.html) has links to resources on careers in chemical information

Cheminfo (www.indiana.edu/~cheminfo) is designed to help people find and learn how to use chemical information resources on the Internet and elsewhere

Chronicle of Higher Education (www.chronicle.com) lists openings for academic librarians

Cunningham, Anne Marie, and Wendy Wicks. 1992. *Guide to Careers in Abstracting and Indexing.* NFAIS: Philadelphia, PA.

Drexel University offers a master of science with specialization in management of digital information and information/library services—see drexel.edu

ELD-NET (www.englib.cornell.edu/eld/listserv/eldnetfile.html) is an electronic discussion list addressing issues related to or of interest to engineering and related subject area libraries and librarians

History of Chemical Information (www.tu-harburg.de/b/hapke/infohist.htm# chemical)
Indiana University School of Informatics (www.informatics.indiana.edu) offers a master of science in chemical informatics
International Union of Applied and Pure Chemistry (www.iupac.org) explains chemical terminology and nomenclature
NFAIS (nfais.org) serves groups that aggregate, organize, and facilitate access to information
Online Library Science Degree Programs (www.worldwidelearn.com/library-science.htm) offers online and ALA-accredited undergraduate and graduate degree programs in library science, digital-information management, information resources, and library and media services
Rovner, Sophie. 2005. The 21st-Century Chemistry Library. *Chemical and Engineering News,* October 10: 52–53.
University of Chicago provides salary information for professionals in chemistry publishing (www.lib.uchicago.edu/cinf/salary/cinf.html)
Society for Scholarly Publishing (www.sspnet.org) works to advance scholarly publication and communication and the professional development of its members
Special Libraries Association (www.sla.org) has Web pages for the local chapter and the chemistry division
University of Indiana (www.informatics.indiana.edu/academics/chem. asp) offers a master of library science degree with a specialization in chemical information

3 Chemistry and Patents

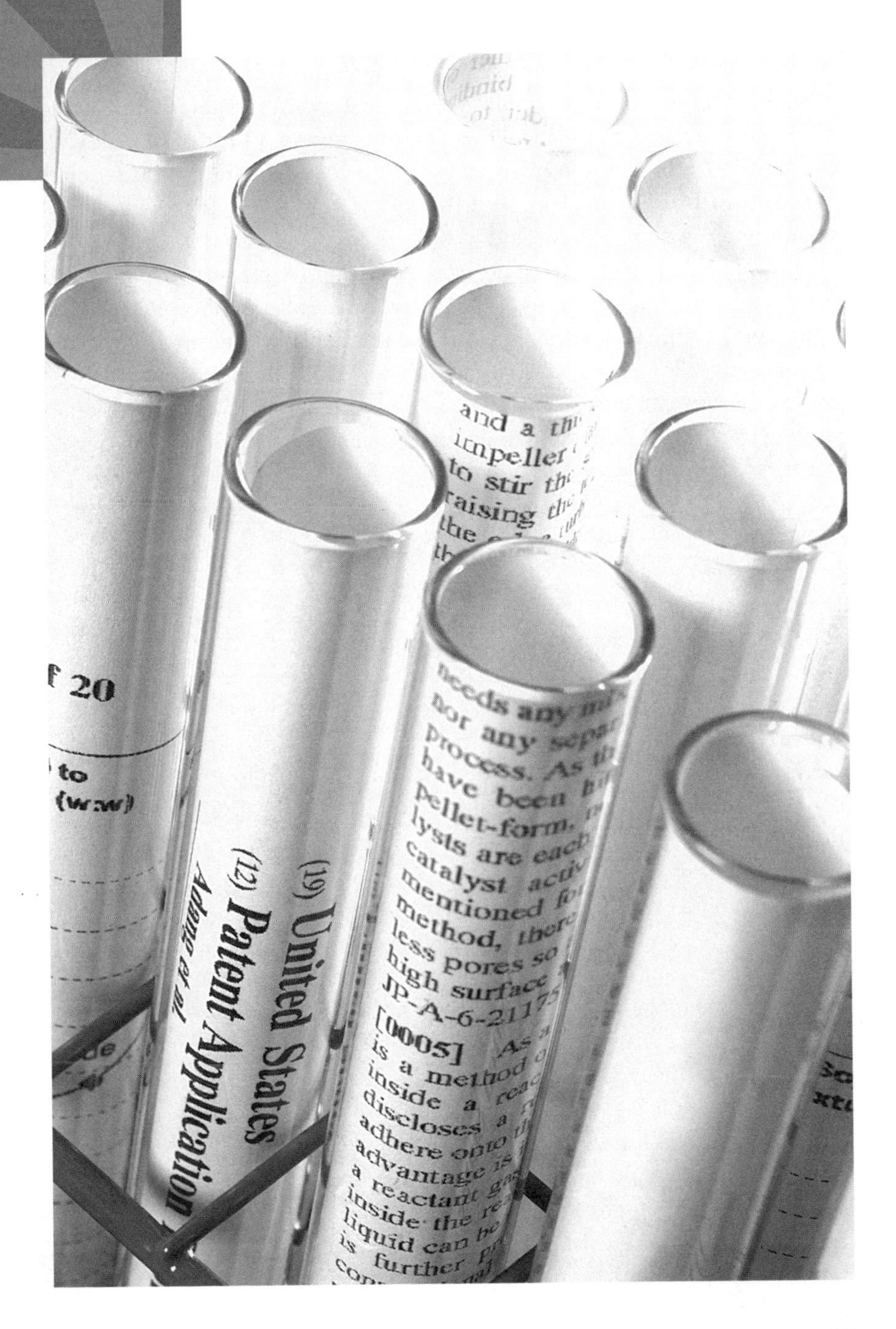

PATENT PROFESSIONALS SEARCH existing literature, develop strategies for the protection of intellectual property, and write and prosecute new patent applications. A career in patent information may be ideal if you enjoy research, logic, and writing, and want to use your chemical expertise outside of the laboratory. There is currently a shortage of people in this field, and many companies are actively recruiting for patent professionals, sometimes even while laying off in other areas. Patent work allows you to continue your involvement with cutting-edge technology, in many cases even before it's published. It requires the ability to analyze complex issues quickly and completely, to work as part of a team, and to write persuasively and well.

Entry into the patent field requires only a bachelor's degree in a science-related field, and many people continue their careers by learning on the job. However, significant advancement requires additional training and education, which can be obtained while working or by going back to school full-time. Salaries increase commensurate with advanced education and experience. Patent professionals work in law firms, industry and universities, and even run their own businesses.

* * *

IF you're interested in a career in the patent field, the first thing to learn is exactly what a patent is, and what it is not. A patent is a particular form of intellectual property (IP), which gives its owner (usually the inventor) the legal and exclusive right to make, use, and sell an invention for a limited period of time. Other types of intellectual property include copyrights, trademarks, and trade secrets. An inventor must apply for a patent, and if valid it is granted by a country's government in return for a full disclosure of the details of that invention. Patents are only granted for inventions that are novel, non-obvious and useful, so close scrutiny of all existing related knowledge (referred to as "prior art") and careful wording of the application are crucial for the success of a patent application. A patent includes two main parts: a description of the invention and a series of claims. The description must contain enough detail that someone else skilled in the art could reproduce the invention from that information alone. The claims at the end of the patent enumerate exactly what is covered, and must meet the criteria for patentability, while at the same time providing the broadest possible amount of protection for the inventor.

Companies of all sizes need people not only to draft and prosecute patents on their own inventions but to develop and execute strategies for licensing other company's inventions, as well as for comprehensive and cost-effective protection of all their own intellectual property. Companies need to know what has already been patented, and what has not, in order to:

- Determine their "freedom to operate"—will planned products or processes infringe on existing patents, or will they be patentable themselves?
- Identify valuable technology assets to be developed or marketed.
- Develop intellectual property strategies—decide what should be developed internally, and what should be licensed from other companies.
- Decide which technologies should (or should not) be patented.
- Write new patent applications.
- Argue for patent applications that were initially denied.
- Protect granted patents from infringement.

There are a number of ways to get started in the patent field, many of which do not require any special training other than a scientific background and an interest in intellectual property. Over time, additional education and training can lead to new opportunities and significant advancement, as well as higher salaries. Salaries for entry-level positions are slightly less than those of a laboratory chemist, but slightly more than for other types of chemical information specialists. "Chemical information positions in general do not command the salaries they should," says Andrew Berks, PhD, who is a senior patent agent at Ivax Pharmaceuticals and was chairman of the chemical information division of the American Chemical Society. He adds, "There are not enough people who know the tools to fill the available jobs, and demand exceeds supply. This should drive salaries up, but it may take 5 to 10 years. As the scientific literature landscape becomes ever more complicated, these jobs will get more respect and I do believe salaries will catch up."

Patent professionals practice their trade in a variety of settings, including law firms, technology-transfer offices at universities, the U.S. Patent and Trademark Office (USPTO), and all types and sizes of industrial settings. Working in this field allows one to explore many different technical areas—sometimes during the course of a single day—and to keep up with the most-current technological advances. The patent professional spends time talking about the latest developments in science and technology with logic-oriented people.

Patent work can also have an international aspect. A given patent protects an invention only in the country in which it is filed, so multiple filings are required for international protection. Timing is crucial—filing a patent application in one country can invalidate later filings in other countries.

The specific types of positions in this field are described below. While job titles may vary, the tasks described are typical for a large number of potential employers.

Patent Examiner

When an inventor wants to patent a new invention, he or she prepares a patent application and submits it to the USPTO in Washington, DC (or the appropriate office in another country). Patent examiners are the government employees who evaluate those patent applications and determine whether or not the patent will be issued.

To become a patent examiner, one must have a background in science or a design-related field. Examiners often have degrees in electrical, mechanical, or civil engineering; physics; computer science; biology; chemistry; or biochemistry. Newly hired examiners receive 200 hours of training during their first year and gain expertise in the entire course of patent prosecution, from initial filing to final issue or abandonment. Patent examiners continually apply scientific knowledge and analytical reasoning, corresponding with patent attorneys, as the process progresses. Appeals for initially rejected patent applications can require many written arguments before a final disposition is reached.

Most of this work is confidential and cannot be taken home, which has both advantages and disadvantages. Since patent examiners work at the USPTO, living in the Washington, DC, area is a requirement.

A position as a patent examiner is a good way to get started in this field, and there are many places to go from there. One possibility is to become a registered patent agent, which requires completing four years as a patent examiner, or passing the patent bar. From the patent office, it is possible to move into a variety of other government offices, including the Office of the Solicitor, the Board of Patent Appeals and Interferences, and the Office of Legislative and International Affairs. Patent examiners often leave government service and move to the private sector, where private legal firms, corporations, and educational institutions frequently need people with expertise in patent prosecution.

The USPTO provides tuition assistance for examiners who want to attend law school at night while obtaining valuable work experience. Attending law school this way generally takes about four years, at which time passing the bar qualifies one as a patent attorney.

Patent Searcher

Becoming a patent searcher is another excellent way to get started in this field. Patent searchers are a specialized subset of chemical information professionals and generally work for large pharmaceutical, oil, or chemical companies that have significant research operations. This type of work requires a feel for computers, an understanding of searching logic and strategies, attention to detail, and an ability to analyze and summarize complex issues. Patent searchers scour the existing literature and prevent companies from wasting work on projects that are not patentable—or worse yet, ideas that have already been patented by someone else. The work is conducted outside the laboratory yet requires a technical background and expertise in formulating research queries. It often requires close collaboration with laboratory personnel to understand what they really want to know—which is often quite different from what they ask for.

Typical questions asked of a patent searcher include "Is this structure known, or are syntheses of any similar compounds known?" and "What companies are active in this area?" In a more general chemical information position, searching may involve more than just patent information, and working closely with both lab scientists and managers. If patent searching is the primary focus of the position, patent counsel would probably also be close coworkers.

Using their knowledge of chemistry, patent searchers scour the existing patent literature for prior art that might make a particular line of research unpatentable, and therefore undesirable, and for inventions that can be licensed and used by the company. They need to be able to read between the lines and identify what was left out of existing patents, and figure out why. The objective is to perform skilled, complex, and expensive literature work to help guide future research.

Search results must be communicated with precision, clarity, and logic. The logical-thinking and problem-solving skills useful in scientific research come in handy here as well. The major difference is that patent analysis is abstract, so there are rarely any completely right or wrong answers. To work in this field, patent professionals must be able to embrace and exploit that ambiguity.

Patent-searching software and tools are very sophisticated, so many larger companies require employees who are experts in their use. In fact, the complexity of modern searching tools means that only skilled practitioners can use them efficiently. There is ample low-cost or free training for most of the specific tools. While there are a few successful independent patent searchers, such as Suzanne Robins of Patent Information Services, Inc., Suzanne feels that "it is imperative that you start out with a company to receive the training and go independent only after many years of corporate experience." Smaller companies may occasionally use consultants to conduct patent searches for them, but they want experts with many years of experience conducting patent searches in industry.

Patent searchers may be promoted up the corporate-management ladder, or movement into licensing or regulatory affairs. Some people prefer to remain patent searchers for decades, whereas others take the test to become a registered patent agent, or go to law school and become patent attorneys.

Patent Liaison or Technical Specialist

Patent liaisons and technical specialists work with both scientists and patent attorneys, gathering data and writing up background material and experimental examples for incorporation into patent applications. This is another excellent entry-level position, where existing chemical expertise can be leveraged to gain experience in the patent process. These types of positions move closer to patent attorney and further away from chemical information.

Technical specialists assist in documenting new scientific inventions, drafting patent applications, and writing revisions and responses to patent examiners during the patent appeals process. They work closely with the inventor to understand the specifics of the invention.

They can also track the IP movements of competitors and help determine IP filing strategies. However, they cannot sign legal documents or communicate directly with the USPTO. They must work closely with a registered patent agent or patent attorney who reviews everything they do and signs off on the official submissions.

Drafting patent applications is a very exacting process. The specific words chosen both define and limit the scope of the invention for the lifetime of the patent. The patent application must explain how the new discovery differs from everything previously known, while assuming that the audience has no knowledge of the specific topic. The most important aspects of the invention must be identified—often from only a sketchy outline provided by the inventor. Analyses must be presented in light of statutory law, case law, science, and prior experience with clients and the patent office. A new patent cannot interfere with an existing patent, but the goal is to claim as much intellectual property as possible for the inventor. Although the writing itself is a solitary endeavor, the subject matter does require teamwork.

Technical specialist positions are often held by those attending law school to become patent attorneys.

Registered Patent Agent

Only registered patent agents and patent attorneys can communicate directly with the USPTO. These individuals have verified expertise in patents. Patent agents negotiate and draft patent agreements, prepare documents for filing and processing, and submit them to the USPTO. They work with legal and product-development teams to ensure proper filing of patent applications and argue the merits of the invention with patent examiners, normally in writing.

Becoming a registered patent agent requires passing the six-hour patent bar examination. A bachelor's degree in one of the following subject areas is required to sit for the exam.

- Biology
- Biochemistry
- Botany
- Computer science (computer science commission accreditation)
- Electronics technology
- Engineering (most varieties)
- Food technology
- General chemistry
- Marine technology
- Microbiology
- Molecular biology
- Organic chemistry
- Pharmacology
- Physics
- Textile technology

There are equivalency options for people who have foreign degrees and work experience. Since only about half of those who take the exam pass it, studying the rules of practice of the patent office before taking the exam is crucial.

A registered patent agent may draft and file patent applications and conduct business with the USPTO but may not practice law anywhere else and may not provide legal advice. Once a patent has been issued, a patent agent may track the life of the patent but may not represent the patent, assignee, or inventor in court, should problems arise. For example, they may not draw up contracts to license the patented technology or prosecute or defend patent-infringement cases.

The next step up the professional ladder is patent attorney, which requires a return to school—in this case, law school.

Intellectual Property Attorney or Patent Attorney

A patent attorney must have all the qualifications of a registered patent agent (having passed the patent bar) and additionally must have a law degree, pass a state bar exam, and be licensed to practice law in at least one state. Law school generally takes three years to complete full-time, or four years part-time.

A patent attorney is a lawyer with special qualifications in the patent field. In addition to drafting, filing, and arguing for patent applications, patent attorneys may represent the patent's owner, assignee, or inventor in court. A patent attorney can prosecute patent applications, provide support for patent litigation, and provide client counseling on various patent issues. Patent prosecution is usually an ongoing, written dialogue between an examiner at the Patent Office and the attorney (as spokesperson for the inventor) about the patentability of the invention. Litigation involves filing and defending patent-infringement lawsuits in court; including preparing memoranda, motions, and briefs; assisting in discovery; and developing strategies for attacking the lawsuit. Client counseling can include preparing opinions on patent validity and infringement claims, negotiating licensing agreements, and devising a patent portfolio strategy that is suited to the client's business objectives, needs, and resources.

Patent attorneys can provide legal advice on strategy—is patenting a new invention the best way to protect it, or would keeping it a trade secret be less costly or more effective? Should the patent be filed now, or should the invention or formula be held as a trade secret for a while to extend the protection? Who should be included as inventors? What federal regulations will a potential new product have to comply with, and will the cost of compliance be prohibitive for its development? All these can be crucial decisions. A little legal expense and expertise at the beginning of a project can save problems and anguish later on.

Patent attorneys may work in any number of environments, including general-practice law firms or boutique firms specializing in intellectual property, at the USPTO or other government agencies, at large corporations as in-house counsel, or as independent contractors. They probably have the most flexibility of any of the patent-related careers. The circle of patent attorneys is relatively small, so many familiar faces appear at annual conferences and other professional meetings.

For patent attorneys who work at a law firm, the concept of "billable hours" plays a large role in daily work life. Depending on the position and the company, they can be expected to work between 1,700 and 2,200 billable hours per year. This means actual productive work, not including breaks, reading literature, managing subordinates, and so on. Keeping track of where every minute is spent, and realizing that someone is paying for that time, is difficult for some people. Realizing that only 50% to 75% of the time spent in the office is actually billable can come as quite a shock. To a certain extent, these long hours are also typical of other patent-information professionals who work at law firms.

Patent law is a competitive service business, so keeping clients happy is essential. For some, being responsible for the well-being of another person's company and being on call at all times can be a huge burden. For many, the challenge and thrill of working on cutting-edge technology makes it all worthwhile.

Profiles

Kerryn A. Brandt

Senior Information Scientist
Rohm and Haas Company

BS, Chemistry Rensselaer Polytechnic Institute, 1979
MS, Organic Chemistry, University of Michigan, 1980
PhD, Macromolecular Science and Engineering and Chemistry, University of Michigan, 1984
MLS, University of Maryland, 1989

Current Position

As part of a team in the company's Knowledge Center, Kerryn Brandt is responsible for providing a full range of technical and patent-search and -analysis services to employees worldwide in support of intellectual property strategy, research, product development, and legal processes. Basically, he is a search specialist, doing patent-related work in support of product-development teams. He also assists IP strategy teams, to ensure that their ideas for changes to processes will not infringe on anyone else's patent, to help assess the novelty of inventions, and to help the teams understand the IP "landscape" in various technology areas. He is also a member of the Knowledge Center's intranet site team—he provides input on which resources, information, and tools should be made available to all company employees and how those resources should be presented to make them usable. Although most of his work is in support of specific project teams, like most information professionals, Kerryn spends some of his time teaching other employees how to use information resources. This can be one-on-one, in small-group settings, or in formal employee-training programs.

Career Path

Kerryn received a BS in chemistry from Rensselaer Polytechnic Institute. At the end of his undergraduate career, the only options he was aware of were graduate school or an industrial position. He recalls, "I had never heard of a career in information science. That was just not on my radar screen as a possibility. This has changed somewhat in the intervening years but not as much as we might like."

He chose graduate school and went on to earn MS and PhD degrees in chemistry and macromolecular science and engineering from the University of Michigan. "One thing graduate school taught me is that I never want to work on the same project for five years again—perhaps I have a short attention span," he jokes.

After earning his PhD, he joined Procter and Gamble in Cincinnati as a staff scientist. He spent three years there working at the bench, improving a super-absorbent polymer for use in diapers. Along the way, he says, "I got the feeling I liked doing the background research better then the actual lab work. I wanted to try to find a way to do what I liked best as the main part of my job.

"Looking back, I might have been happier had I made the move to information sciences sooner, but I needed the experience, so it may not have been possible."

In 1987 he moved to Baltimore where his wife had found a new job. In January of 1988, he started working on a master of library science (MLS) degree at the University of Maryland. This program took him 18 months to complete while working half-time.

A colleague in one of his MLS classes mentioned that a position would soon be opening up at Johns Hopkins. He was able make contact early in the hiring process and was selected for the position in personal-information management for basic sciences at the Welch Medical Library. Initially, he spent a small amount of time sitting at a reference desk and answering questions for library patrons. The main focus of his job was working with faculty, graduate students, and postdocs, helping them find and organize information for their research. "I spent lots of time teaching people how to use resources—one-on-one, and in small groups and formal classes." He was eventually promoted to a faculty position (assistant professor for biomedical information sciences). The position allowed him to get external funding and publish papers on information retrieval, scientific databases, and Web-based distance learning. He also provided patent-search services for faculty and technology-licensing staff.

In 1989, Johns Hopkins received a large government contract to develop the Genome Database, and Kerryn transferred to that project as training coordinator in 1990 to teach people how to use the new database. This was an international collaboration and involved frequent travel, both domestically and internationally, to provide both training and user support. As a direct result of having so much personal contact with end users, he could provide the developers with users' valuable feedback.

In mid-1993, Kerryn's life changed again. His daughter was born, and he decided to cut back on travel and spend more time at home. He was able to return to the Welch Library, where his responsibilities expanded to include providing support for the Genome Database for people at Johns Hopkins, in addition to his previous roles.

Early in 1997, he decided he would like to use his chemistry background more often, instead of handling only the odd chemistry requests that came along. More importantly, Kerryn realized that he "wanted to do more doing and less teaching" and to learn more about patent information, perhaps by moving from academia to industry. Kerryn recalls that he "was open to possibilities but was not mailing out resumes." As he explored his options, he happened to see an advertisement for a chemical-information position at Rohm and Haas, and he applied. It turned out he had met the hiring manager at a previous American Chemical Society meeting, which helped him get the interview. This time, his wife followed him, and in August 1997 he started at Rohm and Haas, where he has been ever since.

In his current position, Kerryn's polymer background is highly valuable, and he uses much more of his chemistry expertise. He works closely with research scientists from all areas, finding the information they need. He might meet with a group of colleagues for an hour to help them define their query and make sure he understands what information they really need. Then he might spend three days alone at a computer, finding that information and analyzing it. This is the detail-oriented, solitary, "heads-down work" that the job requires. He is also able to travel to professional meetings and give presentations. He interacts daily with scientists and engineers and rarely with businesspeople.

This position provides tremendous diversity in the type of work he does—Kerryn researches many different types of polymers, in addition to other types of products. He works with a wide variety of projects and types of people. In very large companies with big information departments, there will be specialists in a specific area, but at midsize and smaller companies, each information scientist works on many different topics, sometimes all in the same day.

Kerryn notes, "The information-science field is growing tremendously in terms of opportunities—people don't need help finding the routine stuff anymore, they now need help finding the hard stuff and interpreting the results they get. The level of queries that scientists are asking are more complex now, which keeps the job interesting and lots of fun.

"In my current position, polymer chemistry is a very useful part of my background, because that's a lot of what Rohm and Haas does. It allows me to understand what I find, because I speak the same language as the research scientists I'm working with."

Another crucial skill, he says, is "the ability to conduct a proper reference interview—to find out what someone really needs to know, which is often not what they asked for." He adds, "In many cases, talking to me helps define the problem. I have lots of interactive meetings using databases and a projector—we can slice and dice the data interactively and in different ways in real time until we match what is available with what they need to know. I also have lots of phone and computer tie-in contact with people in various locations. We don't use video conferencing much yet."

Salaries in this field are good. "I'm pretty happy compared to the Chemical and Engineering News data for chemists with the same amount of experience. If I had taken a managerial track or stayed in the lab, I might have made more money, but I wouldn't have been as happy."

Predictions

The current challenge in this field is information overload—and trying to handle increasingly large sets of information. "We are learning how to do data mining to extract useful information from the vast amount of raw data that is being generated. I predict that data mining, and tools to build information maps of thousands of patents will become more available, and there will be better analytical techniques."

Another challenge is finding the balance between putting search tools on the desks of all researchers, and providing expert search service for the researchers. "It's certainly more efficient for scientists to be able to do their own searches—and they can, for straightforward ones. However, more-complex queries require some degree of expertise and a familiarity with all the available tools. Knowledge of data sources is crucial, because there are multiple classification systems, detailed indexing systems. . . . You really need to use them on a regular basis to use them efficiently."

Andrew Berks

Senior Patent Agent
Ivax Pharmaceuticals

BA, Chemistry, University of California, Santa Cruz, 1980
PhD, Chemistry, University of Colorado, Boulder, 1988
JD, Fordham University, anticipated 2007

Current Position

Andy Berks is responsible for reviewing patent and legal issues for generic-drug development projects. He spends the majority of his time talking, reading, thinking, and writing. Since he works for a generic-drug company, he is often called upon to analyze the patent landscape for a particular drug, to determine if there is a legal market entry point for a generic version. The Food and Drug Administration's (FDA) Orange Book is the official register of approved pharmaceutical products, providing notice to generic-drug makers of name-brand-patent rights, including active ingredients, drug formulations and compositions, and approved uses. Andy evaluates all relevant patents and publications to ensure that a new product his company is planning to launch is not going to infringe on any existing patents. However, if his company believes the patent on a particular drug is invalid or that they can market a competing drug without infringing on that patent, they can file a "paragraph IV certification," indicating the company's intent and justifying their belief that they can market a patented product. "This is a big legal area," Andy says, "because it's basically an invitation to be sued for patent infringement. A lot of litigation takes place around it." He also analyzes infringement issues for drug-manufacturing proposals—all raw materials are purchased from suppliers who must be approved.

Every day he works with research managers and people working on drug-development projects, including marketing and regulatory affairs personnel and project managers.

Career Path

After graduate school, Andy took an academic postdoctoral position. However, he was unhappy in it because he felt he needed more postdoc experience to get a good laboratory job, but, he reflects, "I was impatient to get a real job at that point." He did a lot of library-research projects in those days and enjoyed that more than the laboratory work itself. In the course of exploring options, he submitted a resume to an ACS Meeting Career Exposition and indicated an interest in literature searching, patents, and information science. Hiring managers from American Cyanamid–Lederle Labs were looking for someone to do this type of work and recruited him to become an information scientist.

"Without that first step, I'm sure I would be in a very different place now," Andy says. "Moving to a searching and literature function was a big step, and the kind of thing they don't tell you about in grad school. The reality is there is a big need for scientists, including PhDs, in many functions outside the lab. For example, in drug companies a lot of lab scientists move into regulatory affairs and clinical research. I was recruited into a niche where I had both the talent and the science background needed for the job, and found I really enjoyed the work."

Within a few years, Andy became an internationally recognized expert in patent searching and information. After doing this type of work for about 12 years, he felt a desire to upgrade his career and seek new challenges. In investigating his options, he realized that since searching work is relatively specialized, he would have to change organizations to advance, which often means relocating. Andy decided that, for him, the next logical step was patent law. He found a position as a patent agent in the legal department of a major generic-drug company and started attending law school at night. His next career goal is to finish law school and pass the bar, making him a patent attorney.

He recalls, "I came into contact with patent lawyers early on in my searching career, and when I felt an urge to move on, law seemed like a logical choice. It was an 'I can do that!' feeling. Patent lawyers have greater responsibility and more powerful positions than patent agents, so it should be a good move. As an attorney, I also expect to have much greater flexibility about where I can work." As an added benefit, while patent searchers are paid about the same as other laboratory workers, the salaries are significantly higher for patent attorneys.

Eventually, Andy may become a partner in a law firm, an independent practitioner, or general counsel for a corporation.

Predictions

Andy says, "The field of patent law will continue to grow. Patents, and other forms of intellectual property protection, are more important than ever and only increasing in importance. In addition to a science background, people who want to move into this area will find computer skills, writing skills, and analytical thinking ability crucial. The hands-on experience I got in school has been indispensable, especially my background in organic synthesis. I believe synthetic chemists can do it all, because they need to be generalists in all areas of chemistry. I don't think many other branches of chemistry can say that."

Alex Andrus

Patent Counsel
Genentech, Inc.

BS, Chemistry, University of California, Los Angeles, 1976
PhD, Chemistry, University of California, Santa Cruz, 1980
J.D., University of San Francisco, 2002

Current Position

Alex Andrus is an attorney in the legal department of a pharmaceutical company. His position requires prosecuting patents and managing contracts and collaborations involving intellectual property. The company's scientists are his clients. Several times per week he meets with them formally, either in their offices or labs, or in his office or a conference room. He attends research seminars and visits the labs. He talks with the scientists about ideas that may be patentable, presentations they want to make, manuscripts they want to publish, interactions with collaborators, and their obligations to the company as an employee.

Career Path

Alex's first job after graduate school was at Merck & Co. There were hundreds of synthetic organic chemists, most of whom had been at Merck their entire careers. "It was a great experience, but I quickly developed a sense that I didn't want to be shaking a separatory funnel when I was 60 years old," he says. After 17 years in a traditional laboratory career, he found himself at an instrument-manufacturing company as the leader of a group developing oligonucleotide synthesis products. The company he worked for held the rights to dominating patents and granted licenses to other synthesizer and reagent manufacturers. Some competitors refused to take a license, and his company filed a patent-infringement lawsuit. The other company made the expected counterclaims and defenses. Alex became involved with the lawsuit as the in-house expert, working with the legal department and outside attorneys. It quickly consumed a significant amount of his time, but he realized he was enjoying both the process and the players.

He received encouragement from the company's legal department to become more involved in patent work, this time with an inventorship dispute with a university. A university research group started a company to commercialize technology that Alex's company thought they had contributed to through an informal collaboration. Although the lawsuit was eventually settled, Alex learned the importance of documenting ideas properly and having agreements in place. He says, "Scientists are wonderfully creative and eager to share information. However, they need people like me around to preserve their rights and keep them out of trouble. This is not altruistic, of course, because employees usually must assign their rights to the companies for which they work."

At about this time, Alex was asked if he would like to make a career change, transferring from research to the legal department. He would work for the company in the patent craft full-time during the day, and they would pay for him to attend law school and prepare for the patent bar exam at night.

He recalls, "Going to law school was not something I expected to do. I never aspired to be an attorney, and it never had any appeal for me. I wasn't even interested in television shows or movies about lawyers! The practice of patent law is different; it's a niche that most lawyers know nothing about. When Applied Biosystems offered to pay for law school and accommodate my work schedule, I jumped at it. My motivation stemmed from the challenge and excitement of doing something different while also using, and building on, my chemistry background." Other valuable skills for patent work include interpersonal skills, such as the willingness to seek people out, get them to talk, and build relationships. Being detail-oriented also helps.

This career path is not unusual. Most patent attorneys get their science or engineering education first, then go to law school. Alex was very fortunate to also receive one-on-one mentoring and training from the patent attorneys at Applied Biosystems.

Several factors combined to make this transition possible. Scientific knowledge is paramount—without being an expert in at least one technical area, it is difficult to contribute. Writing skills are also critical. The only question Alex was asked before being offered the new position was: "Do you like to write?" Patent work requires the ability to create coherent descriptions of someone else's ideas and experimental data. Patent applications are written in a specific but unusual style, and Alex finds the most difficult part of writing is putting a concept into words on paper for the first time.

Since graduating from law school in 2002, Alex's primary job has been patent prosecution. Prosecution is the entire set of activities necessary to obtain a patent. He starts by investigating an idea, which may or may not turn out to be a patentable invention. He talks with people who have knowledge of the potential invention, usually a scientist who has done experiments and generated interesting data. During a formal interview, he asks a lot of open-ended questions and takes notes or tapes the conversation. He must listen carefully and not presume anything. This begins a long, primarily solitary, process of searching the literature, that is, the "prior art," for relevant disclosure concerning the big issues of patentability and the company's freedom to operate in this area. Sometimes the decision is made not to patent the idea, even if it is patentable, but to instead hold it as a company trade secret.

The most exhausting and creative part of his job is writing, consulting with the inventors on multiple drafts of the patent application before it is filed. The examination process begins with the various patent offices, with the goal of acquiring one or more meaningful patents. This process can include framing persuasive replies to the patent examiners' inevitable rejections of the invention. It is a highly formalized process consisting of a series of compromises and strategic decisions based on the perceived importance of the invention, the research and development activities, the company's confidence that the invention is patentable, and various economic considerations.

A secondary part of his job is to ensure that collaborations with other companies and universities go well by interpreting the agreements and providing guidance.

He mainly interacts with other patent attorneys and agents and the legal department staff at Genentech. Since it's a large company, everyone brings their own scientific background and legal expertise.

He says, "The patent craft—and I call it a craft because it requires mentoring and a certain period of apprenticeship before you can really operate independently—employs attorneys, agents, and people who do background research including literature searching and diligence. When inventors have a good idea and you show them you understand it and speak their language, they really appreciate that. It is very satisfying if you can then deliver a high-quality patent application that represents their thinking and their work.

"After seven years at the patent craft, there is still a lot to learn. Patent rules and practice are like the tax code, constantly changing and forcing you to keep up. Technology changes, too, especially the subject matter I practice in. So, just trying to get better at patent prosecution is career progression for me."

Advice

Alex advises, "Learn your scientific or engineering field really well, get some significant lab or technical experience, write articles to develop your writing skills, and find some patent people to talk to.

"Most lawyers love to talk about themselves, especially the horrors of law school, even when they're not billing you for their time. If your company has in-house patent people, show some interest in their work and get them to talk. Most universities have technology transfer offices with patent attorneys. If there aren't any at your institution, go out and find some and talk to them.

"Like a lot of fields, it's hard to get that first job. To learn the patent craft, you have to find someone who will work with you, one-on-one, review your work, and engage in constructive criticism. Law firms and in-house patent groups do this, but they all recognize how labor-intensive it is so are selective about who they take on."

Predictions

According to Alex, "All indications from industry and governments are that intellectual property will continue to grow in importance. Foreign governments are falling into line, through implementation of the World Trade Organization's Agreement on Trade-Related Aspects of Intellectual Property Rights (TRIPS) agreement, to grant and enforce patent rights. It is now well accepted that the primary value of technology companies is their intellectual property. Patent attorneys are the paranoid gatekeepers of those crown jewels."

Anita L. Meiklejohn

Principal
Fish and Richardson P.C.

BS, Chemistry, Cornell University, 1981
PhD, Biochemistry, University of California, Los Angeles, 1987
JD, Boston College Law School, 1995

Current Position

Anita Meiklejohn is an attorney practicing intellectual property law at a large, national law firm. Her clients are mainly biotechnology companies, pharmaceutical companies, universities, and medical-device companies. She helps them obtain patents for their technologies and avoid infringing the patents of other companies. "During the development of a new drug, for example, I study the patents owned by other parties and advise my client regarding the scope and validity of the patents, so that my client can direct drug development in a manner that avoids infringing the valid patents of other parties," she says. "I also study the published literature and patents to determine whether my client's product is likely to be patentable. I advise clients regarding the licensing of patents and technology, and I help draft and negotiate joint research agreements."

Most of her time is spent writing patents and analytical letters, meeting with clients, and reviewing work done by more junior attorneys. She also spends a portion of each day dealing with matters related to the business side of managing her practice and the firm.

Career Path

Anita came to her law firm directly from a postdoctoral position and worked there while attending law school. She had read about some patent litigation and issues surrounding patents and other types of intellectual property. It sounded intellectually challenging and analytically complex. She met with the technology-transfer manager at the university where she was doing her postdoc and learned that there was tremendous demand in intellectual property law for people with the right technical background. She decided that this was a chance to do something that was interesting and hard, but in a field that seemed to offer more opportunities than academia.

She recalls, "I felt that a legal career would provide both more stability and more flexibility. I had been considering an academic career, but it seemed I might end up spending many years moving from here to there to further my career and that I might have little choice in where I lived. In my current field I can work in any larger city, and in a city the size of Boston there are dozens of options."

Many law firms specializing in intellectual property have a hard time finding lawyers with a PhDs in chemistry or molecular biology, so they hire people straight out of graduate school, post-docs, or industry to be technology specialists. These technical specialists work at the law firm, doing pretty much the same work as new attorneys. After a year or two, a technology specialist will start going to law school full- or part-time while continuing to work at their firm. For the first five years, until she graduated from law school, Anita was a technology specialist.

While this career path may have been typical in the past, people are increasingly going straight from a PhD program to attend law school full-time. Some people go straight to a company and work in the legal department, but it can be difficult to get good training this way unless the legal department is fairly large.

During the time she was in law school but working nearly full-time, Anita had the chance to become deeply involved in litigation on behalf of a large biotechnology company that was a client of the firm. She was one of a small number of people in the firm at that time with the right technical background for that case. She was given a great deal of responsibility, traveled widely to assist with discovery (e.g., document review and depositions), and assisted with two trials. Between school and the litigation she was generally working from 9 A.M. to midnight or later six or seven days a week for months on end. "It was worth it because I learned to work under tremendous pressure, make decisions, and manage a team. I gained a tremendous amount of confidence because I learned that with the right attitude and a good team you can do things that seem impossible," she says.

Upon graduating from law school, Anita became an associate at the firm, and eventually she was made a principal. This means that she, together with the other principals, are the owners of the firm, with the ability to vote on all important firm decisions. She has been working in intellectual property law for nearly 15 years. Her current career goal is to continue to serve her clients well while helping her firm increase its diversity and improve training of young attorneys.

In her current position, she interacts with scientists, lawyers, and managers that work for her clients. She also interacts with lawyers and, to a lesser extent, scientists and managers at companies that her clients are adverse to, for example, in a due-diligence or licensing context. She works closely with other lawyers, technology specialists, paralegals, and the administrative staff at her firm, particularly her invaluable administrative assistant. The only thing she would change about her career path is that she "would have bought a BlackBerry® sooner." She adds, "Seriously, I would not change much, even if I could. I am having fun.

"I enjoy the variety of topics and the variety or people I encounter. I love the challenge of making decisions and providing advice in an area of law that is constantly evolving in the context of technology that is evolving even faster than the law. The most difficult part is being pulled in different directions by different clients. Often, several different clients want your full attention and you have to do your best to allocate your energy."

Advice

Anita advises, "To succeed in this field it is essential to have a strong technical background, but plenty of people have that. People who succeed have sound judgment, the ability to make important decisions using imperfect or incomplete information, the ability to spot and analyze potential problems quickly, people skills, and the ability to write extremely well and relatively quickly.

"If you look around, you will find that you know somebody in this field or you know somebody that knows somebody. Call them or send them an e-mail, and learn about the field from somebody who knows. Think about the demands of the field. You need to be able to write very well and quickly. The writing is not technical writing, although it needs to be both highly analytical and persuasive. These two goals can be at odds with readability, but if your writing isn't understandable and enjoyable to read, it cannot be effective. You also need to be very service-oriented. Clients pay hundreds of dollars an hour for patent attorneys' time, and they expect to get timely, cost-effective work every day."

Predictions

Anita says, "I see a great future for this field. Intellectual property law does not have the up and down cycles you see in corporate law. I think it will be harder to get into a firm without going to law school first, but there are still many good opportunities."

ADDITIONAL RESOURCES

American Bar Association Section of Intellectual Property Law (www.abanet.org/intelprop/home.html)

ACS Division of Chemical Information (www.acscinfiorg) has a worthwhile electronic mailing list

ACS Division of Chemistry and the Law (membership.acs.org/C/CHAL)

American Intellectual Property Law Association (www.aipla.org) is an association of attorneys specializing in the practice of intellectual property law

City University of New York (mail.sci.ccny.cuny.edu/~phibarn/careers.html) posts a salary survey for patent searchers

Dialog (www. dialog.com) periodically runs workshops on patent searching

Patent bar exam (www.uspto.gov/web/offices/dcom/gcounsel/oed)

Patent Information Users Group (PIUG) (www.piug.org) is the premier organization for patent professionals. The Web site has great links and the discussion list is free; many companies advertise patent-searching positions there

Patently-O blog (www.patentlaw.typepad.com) is a great place to learn about issues of interest to intellectual property attorneys

Pharmaceutical Education & Research Institute (www.peri.org) periodically teaches an onsite, two-day course called Patent Information for Pharma/Biotech

Questel (www.questel.orbit.com) offers periodic workshops

Special Libraries Association (www.sla.org)

STN (www.cas.org/stn.html) offers free patent-information workshops periodically

United States Patent and Trademark Office (www.uspto.gov)

4 Chemistry and Sales and Marketing

NOT MANY CHEMISTS consider sales as a potential career path. In fact, many scientists refer to sales as the "dark side" of the chemical industry. However, if you have the right personality and background, a career in technical sales and marketing can be enjoyable, rewarding, and highly lucrative. Technical sales requires not only a scientific background (to understand the technical products and their potential uses) but also an outgoing personality and good interpersonal skills to build relationships with fellow scientists and inspire their trust.

Technical sales is not like used-car sales—customers want facts and details not a sales pitch. Technical-sales people exist to serve the needs of their customers and so must listen and understand their problems then offer products and services that will solve them. "Basically, you solve problems for people," says Ted Gast, the managing director for the Carl F. Gast Company. "If you are application-oriented, this could be the career for you."

Technical sales is an excellent way to move from the laboratory into a more business-oriented career. As you learn the available products and specific market and gain more business experience, the next step along the "dark path" may be marketing (directing the evolution of technical product lines), business development (expanding markets for the company's products), and many aspects of product distribution. This chapter discusses careers that emphasize the understanding the technology and dealing with the end customer; the following chapter discusses careers that have more of a business and corporate emphasis.

* * *

No matter how good a product is, if the right people don't know about it and realize why they need to buy it, the company that makes it will still go broke.

Technical-sales people represent complex products to potential customers and advise them on the benefits of these products, including both quality and performance features. They assist customers in picking the right product for their particular scientific and economic needs. Technical-support people also help customers after the sale, answering questions and providing advice about proper use of the product. Scientific and technical products are not commodities for which low prices rule; rather, they are developed and sold through long-term relationships with intelligent customers. A scientific background is required to understand both the company's product line and the customer's application area, and consequently to explain synergies between them. It is important to develop long-term relationships and trust with customers, to be in touch with current customer needs, and to be able to project what the market will demand in the future. Moving to the business side of chemistry also requires a shift in philosophy from scientific curiosity to financial viability, and a more outgoing personality than the typical laboratory scientist has.

Getting Started

Chemists can learn sales on the job if they have the right personality, but it's very difficult to teach science to someone who has only a business background; interpersonal skills cannot be taught. Many chemical-manufacturing companies hire as many BS chemists for sales positions as they do for laboratory positions (about 60% of all chemical-sales people have a degree in chemistry), and for some of them the first time they consider sales as a career is when the position is offered to them.

An advanced degree is often an advantage, especially for more-complex products, as it provides instant credibility with potential customers. Some companies use "sales scientists," experts in their technical field who travel with the salespeople to talk only about technology, leaving pricing and contract details to the less-technical salespeople.

Obviously, a business education in addition to a science background is the ideal preparation for a technical sales career. While business skills can be learned on the job, a business, marketing, or communications degree, in addition to a science degree, is an unusual (though less so every day) and valuable combination. For people changing careers later in life, many community colleges and university extension programs offer business courses that can be taken at night or on weekends to provide education in the business of science.

A good way to learn about technical sales is to talk to people doing that job. Salespeople who call or visit you can be valuable sources of information about the job and the job market. In addition, tradeshows, expositions, and national meetings are perfect opportunities to talk to salespeople working the booths in the exhibition hall. Those are the technical-sales people and sales scientists, who will be able to explain what the job is really like.

If possible, speak directly with a top account manager at a life-sciences company for an inside look. Taking that person to lunch would be an ideal way to find out more about the field and get some pointers on how to begin a career. Executive recruiters can also be a valuable resource, so create a list and let them know you are interested in joining a company in a sales or technical position.

Customer Support, Inside and Outside Sales

A good way to get started in the business of science is with a position in customer support or inside sales. Customer support involves answering existing customers' questions about a product. Inside sales involves answering inquiries from potential customers and selling products to those who call the company. These positions generally require a technical background in the product field and minimal familiarity with the company's products. In some cases, these positions can also involve cold calling prospective customers and trying to sell the product to them. These positions generally require regular hours, worked in the company's office. Huge amounts of information about the company's products must be read and digested, but one will present only the salient points for each individual customer to that customer.

Providing answers to customer with problems and advising them on proper product use is a great way to learn about a company and its products and to build relationships with customers. Business and communication skills can be learned and practiced. From there, moving within the company to an outside-sales position is often a natural step, which generally allows more flexibility in hours and location, and greater potential income. A little tenacity in pursing advanced opportunities will demonstrate to supervisors that you have the determination necessary to succeed in a sales position.

Outside-sales positions require significant amounts of travel, with up to 80% of the time spent on the road. Outside sales can require a lot of driving, so a clean driving record is important. (Some companies now ask for a driving record as a condition of employment.) Many technical-sales people feel there is a direct correlation between the amount of time spent on the road and sales revenues, and the conventional wisdom is that "if you're at your desk, you're goofing off." Telephone and e-mail are good ways to keep in touch with customers between visits, but most selling is still done face-to-face. The working hours can be flexible but must be arranged to accommodate the customer.

Often, outside-sales people work from their own office, geographically distant from the corporate headquarters, so the ability to work independently is crucial. It is becoming more common to have a home office, which presents its own opportunities and challenges. Home offices have lower overhead and offer greater convenience, but the tendency for business and personal activities to overlap greatly increases, and one must take care to ensure neither sphere overwhelms the other. The tax implications of a home office must also be considered.

Typical tasks for outside-sales people include calling on existing and potential clients; writing quotes, bids, and proposals; reviewing contracts; and helping clients sort out problems. Establishing and maintaining relationships with customers and keeping them informed of new products and product lines, pricing, availability, and delivery issues are also important. Extensive record keeping is essential, to make sure that nothing is missed and that each customer is kept up to date on all products of interest to them.

Selling technical products requires both preparation and perception. On average, 3% of leads from cold calls turn into requests for quotes, and only a third of these quotes turn into orders. That means it can take 100 sales calls to get one order. Learning to handle rejection without taking it personally is key to longevity in this field! Having an outgoing personality and enjoying talking to people can help make those 99 "no's" easier to take. Doing comprehensive research before approaching a prospective client can improve these numbers significantly—the more you know about how their business works, the better able you will be to target your presentation, and the better your chance for a successful sale.

Communication skills are crucial for all sales chemists. They must listen to their customers, for the customer is the expert, and theirs is the only opinion that matters. This requires a different mindset than that of the laboratory researcher, who is used to being the expert on everything.

In many cases, a sales chemist's opinion will be met with skepticism because they are assumed to be biased in favor of their own products. To be successful, salespeople must establish trust and build long-term relationships with their customers. "Nothing gets done unless the customer trusts you—you must meet them on a personal level first," says Trish Ward, formerly a senior account manager for Spotfire, Inc.

Reputation is crucial. Salespeople must represent their company well at all times, without speaking ill of their competitors. Some salespeople take courses or read books on business etiquette. They must be professional and confident at all times—much business is done over meals or in other nontraditional business locations, so they must be comfortable in all situations.

The technical-sales person is the point of contact between the company and both existing and potential customers. As such, salespeople can provide valuable customer feedback to the company by passing along customers' suggestions for improvements and new products they'd like to see. They become the link between the technical development staff and their market. However, salespeople are often in the middle—perched between what the customer wants and what the company can deliver. Diplomacy and tact are required to keep both sides happy.

Character Traits

In addition to being able to meet people and quickly develop a rapport with them, technical-sales people need to have several other important personality traits. First, it helps to have a confident, slightly aggressive personality. The ability to take "no" for an answer but not take it personally can also come in handy. Self-motivation, initiative, and good time-management skills are crucial, especially for independent, outside-sales positions. A positive attitude toward foreign countries and cultures is required, and fluency in another language is a plus. The ability to write easily and well is also important.

Attention to detail is a key skill. Managing a paper trail and organizing records are important ways to help you document and recall exactly what each customer said. Fortunately, chemistry-laboratory note-taking skills provide excellent preparation in documenting details. The ability to organize, prioritize, and follow through are vital, as is the self-confidence to work in an isolated and potentially hostile environment. Salespeople must be able to think on their feet, focus, and not be distracted by extraneous information.

In order to succeed at either inside or outside sales, one must develop the ability to read people. Salespeople have to determine who the key people are—who will make the decisions about what to purchase and who must be convinced of the merits of the product. Reading between the lines and identifying what the customer really wants, not just what they say they want, is crucial. Negotiating deals is an art in itself and can take years to master.

Chemists who have spent a few years in the lab before transitioning into a sales position can truly understand the needs of their customers. They speak the same language, are informed about their customers' needs, can respond proactively to problems, and can report constructively on other innovations in the field. They can build solutions for one customer that can be generalized to benefit multiple customers.

A science background is valuable in that it allows a salesperson to identify trends caused by advances in technology, changing environmental regulations, or the economics of low-cost manufacturing processes or locations. Analytical skills are used to evaluate all these changes and trends and to make appropriate business decisions. Salespeople may need to conduct a market analysis, identify trends, and forecast future sales. Decisions must often be made quickly, before market conditions change and opportunities vanish.

Some companies use "application scientists" or "sales scientists," who provide primarily technical pre-sales support and do not deal with pricing or contract issues. The salespeople call them in to assist with the technical sales by demonstrating the product and responding to detailed technical or scientific questions. They may travel to customers' sites but have nothing to do with money or making deals. Application scientists often travel with business development people (see chapter 5), who are responsible for "making the deal."

Technical Buyer

Another entry-level position is that of technical buyer. Just as someone from the manufacturer must sell the product, there is someone on the buyer's side responsible for purchasing it.

Technical buyers coordinate purchasing throughout a company, including ordering supplies, balancing the sometimes conflicting needs of quality, availability, and price. Depending on the size of the company, a technical buyer may manage all aspects of the business cycle, from researching product specifications to creating requests for bids, placing purchase orders, tracking delivery, and paying invoices. They will negotiate terms and conditions with established suppliers and identify new and better-performing suppliers. They may also conduct audits.

In many companies, inventories are kept on a just-in-time basis, meaning usage and delivery patterns are carefully monitored to make sure that supplies are available when needed but extra corporate capital is not tied up in inventory.

In some cases, technical buyers procure not only supplies and equipment but also services. This can involve evaluating and hiring contractors to perform specific duties, either for a specific project or for ongoing needs.

Career Development

From technical sales, many people move into marketing, product management (deciding what products should be sold, updated, or discontinued), or sales management. Some salespeople move into management for one of their former customers or become technical consultants (selling a service, such as their expertise on the company's product). Many technical-sales people move on to work for one of their clients, as an expert in the product they formerly sold.

Product management is a traditional path. Product managers take responsibility for a particular product or line and promote and build a reputation and client base. This can be risky—it's great if the product takes off, but the product manager may be blamed if it does not. Another way to move is up to become a sales manager—the person who manages the sales representatives. Moving up the corporate ladder usually means more contact within the company and less contact with customers. While outside-sales people may work out of a home office, product-management and sales-management positions are more likely to be based in the corporate office.

However, these changes require additional expertise. Even from the laboratory, chemists are often promoted into management and budgetary positions for which they have no formal training. The further one moves into the business side of things, the more likely it is that a formal business education will be required. Business classes teach fundamental business logic, such as how to evaluate the return on investment (ROI) or the cost-to-benefit ratio of a particular course of action. A master in business administration (MBA) degree may be required for marketing or for more-senior positions. "While an MBA can help you learn how to ask questions from a business perspective, it is not a magic piece of paper that will guarantee a high salary. In a purely financial sense, only an MBA from a good school, obtained early in your career, will return enough income increase to pay for itself," says Dave Hartsough, PhD, MBA, and director of Informatics and Modeling for ArQule, Inc.

Management and marketing positions use a wide variety of terms in their job titles, including technical sales and sales management; product and market research, development, and management; direct sales, technical service, market communications, customer service, distribution, and many others. The specific responsibilities for a given position will vary with the size and structure of the company. As a general rule, the smaller the company is, the more varied the responsibilities will be for a single position. In any of these positions, interaction with both customers and salespeople and an understanding of the problems and needs of the marketplace are the main focus.

Marketing activities have both inbound and outbound components. Inbound marketing focuses on the product itself and product-management activities, such as converting the market needs, customer requests, and product enhancements into product requirements, and includes the day-to-day activities that ensure that the product that is developed is what the market wants. Outbound marketing activities include preparing data sheets and technical or marketing collateral for the customers participating in tradeshows, advertisements, seminar series, and so on, that focus on the customer.

Working in a sales position often has a hidden fringe benefit. Technical-sales people, by definition, interact with a wide variety of people and so have a huge network to tap into when they're ready to try something new. This is especially important since the higher up a sales position is in an organization, the less likely it will be advertised—these positions are most often communicated by word of mouth or filled by specialized recruiters.

Compensation

Salespeople are some of the highest-paid people in a company because their value is easily measured in terms of the dollars they bring in. As a rule of thumb, a salesperson should bring in product sales of 10 to 20 times their gross salary (including benefits and perks) each year. Since their value is so apparent, sales positions are often the last to be cut when budgets are tight. However, it is easy to judge salespeople only on the dollar value of the sales they close. Quotas become a motivating, even demoralizing, factor, because results are required for longevity and commissions, but outside circumstances affect one's ability to meet goals. Furthermore, salespeople are no longer working on scientific research projects—negative results do not count, and the timetable for achieving results is much shorter.

Salespeople can be paid in a variety of ways—or in a combination of ways. When negotiating a job offer, it is essential to understand how the compensation package is structured and to realize it may take many months before one makes a sale. Compensation may include any of the following components:

- Salary is a dependable, fixed amount paid on a regular basis, and increased perhaps yearly.
- Commission is a percentage of the revenue brought in that is returned to the salesperson. Commission-only positions should be avoided except by seasoned salespeople. Newer salespeople may get a base salary with a commission added on.
- Some companies may offer a draw, which is basically a loan against future income and must be paid back.
- A bonus is a lump sum that is paid as a reward for exceeding a sales quota, landing a big account, or accomplishing some other prespecified goal.

Some companies also offer extra perquisites, such a company car, credit card, cell phone, laptop computer, and so on. Most companies will pay for reasonable expenses while the salesperson is on the road for business, but many companies are drawing a hard line on what is reasonable.

In any case, most good sales professionals will probably end up earning more than they would have had they stayed in the lab.

Profiles

Alan Gregory Wall

Technical Service Representative
Sigma-Aldrich Corporation

BA, Biology and Chemistry, University of Missouri, St. Louis, 1974
BS, Education, Biology, Chemistry, and General Science, University of Missouri, St. Louis, 1976

MS, Chemistry, University of Missouri, St. Louis, 1981
PhD, Organic Chemistry, State University of New York, 1985

Current Position

Greg Wall is a technical-service representative, which means he (along with 25 other technical-service representatives) handles the inquiries for more than 80,000 chemicals and approximately 30,000 lab products offered by the Sigma-Aldrich Corporation. Greg is often asked to provide information to customers regarding products' chemical and/or physical properties, to make recommendations for product use, to troubleshoot applications, and to advise on the safe handling of a particular chemical, its proper disposal, and much more.

Most of Greg's time is spent being available for telephone consultations or answering e-mail inquiries as they come in. He answers questions for teachers, academic and industrial researchers, lab technicians, purchasing agents, doctors, lawyers, customs agents, freight shippers, and colleagues.

Some queries are simple and can be answered right away; others require more in-depth research. Basic requests from customers include requests for product-certification, information, or material-safety data sheets; information on the application of a product; documents about a product's quality; and advice on how to navigate the corporate Web site to use the vast technical resources available there.

Greg says, "Through the years I have been asked many an odd or strange question. I once received a call from a Lexus dealer in Florida who had just received a muffler from California, and in the box was a bottle of one of our compounds. They wanted to know what it was for. I asked if the box it came in looked damaged. It was. I advised that apparently their box and our box got damaged and everything was put into one box. I had the material picked up and returned."

"Another call involved the use of an antibody that worked in most applications but not in all. It was observed that in an MES buffer, the antibody–antigen interaction did not occur. How can you explain this? We first thought it was a trace impurity in the buffer. This required a thorough review of the raw materials used, our production method, and our analytical methods. Further experimentation determined that the phenomenon was not observed at all concentrations of the buffer. For some reason, there was masking of the antigen or the antibody's epitope by this buffer that was concentration-dependent. The problem was solved by changing the buffer."

To find answers, Greg can access a corporate database of product information. When he has to research a new answer, he enters that into the database as well. After review, the information gets posted on the company's internal Web site and is viewable by technical-service representatives worldwide. Sometimes this information gets incorporated into a data sheet for the product, which is then posted on the company's public Web site.

Between calls, Greg works on special projects, such as proofing the next version of a catalog, improving the Web site, and evaluating new product suggestions.

Career Path

After receiving a BA in biology and chemistry, Greg was not sure how to apply what he had learned to the work world. As an undergraduate, he taught labs and did research in biology. After graduating, he did some substitute teaching. He enjoyed doing this and ended up going back to school to get his teaching certificate; he then got a job teaching chemistry at a public high school.

However, after teaching public school for five years, Greg was making less than $12,000 a year and had started a family. He decided the only way to get ahead was to further his professional career and get an advanced degree. Greg returned to school to complete his graduate work in organic chemistry.

He says, "I think that if I had realized and explored my potential as a chemist earlier in my career, my professional experience would have had a different outcome. As opportunities presented themselves, I should have investigated and considered them as possibilities for professional growth. I now take opportunities that come along more seriously."

Normally, employers prefer that one has a couple of years in a postdoctoral position before going into an industrial position, but Greg chose to immediately pursue a career in industry. He sent out 20 letters to various chemical companies, looking for any type of chemical job available in the St. Louis area, where his family was. Greg got 14 responses and 2 interviews. He accepted the position in technical service at the Sigma-Aldrich Corporation and has been there ever since.

Greg chose technical service because he had "always enjoyed learning and teaching biology and chemistry." He says, "The product line offered by Sigma-Aldrich complements both disciplines, and it allowed me to use my knowledge in these areas to answer questions, and to learn new methodologies from other sciences and apply them into new areas."

Greg feels that choosing to become a chemist in technical service has allowed him to make contributions to the scientific community and to continue to apply chemistry to the sciences while exploring an ever-evolving world. His position requires a working knowledge of biology, biochemistry, chemistry, immunology, molecular biology, and tissue culture, in addition to computer skills. He says, "The thing I use the most is my knowledge of general and synthetic chemistry. This has allowed me to understand the processes that go on in a biological world and take it down to the level of making and/or breaking a chemical bond."

While salaries in technical service may be less than those in industrial-product development, the job responsibilities are different. Greg points out, "Even though you may have the same educational capabilities, when you are expected to make a product that generates money, it is a lot different from educating someone about the use of your products."

Greg has had his share of unexpected opportunities. He says, "The more you give of yourself, the more opportunities you will have. For example, a while ago I was at a local ACS meeting, and one of my colleagues asked if I ever thought about going back to teaching. I expressed an interest in this as a possible career transition. I was almost immediately offered a position as an adjunct professor of chemistry, and I seized the opportunity. I am enjoying getting back to my teaching roots and exploring this as a next career.

"One of the biggest challenges in this field is to realize that I am on a continuous learning curve in all areas of the sciences, and it never plateaus. As business interest in the sciences expands, it is necessary to gain immediate expertise in these new-initiative areas with minimal formal training. The greatest scientific opportunity is to be able to apply my knowledge and understanding of a molecular process to a researcher's application and solve their problem."

For the future, the greatest job opportunities would probably come either from internal job advancement or from the exploration of new opportunities from external networking. Gerg says, "As I approach an age in which I can potentially retire from my current position, I realize that there are many more years ahead of being a chemist, and I can now explore some of my own interests. Over the years of accumulating technical information and applying it, I have seen many possibilities, and I am currently pursuing some of them.

"I have always wanted to have a consulting business that would allow me to use my scientific and educational background. I believe there is a need in the private and commercial sector for programs that integrate scientific understanding into people's daily lives. . . . I am sure that there are things that have yet to happen that will open other opportunities for me to grow as a chemist."

Advice

Greg advises, "You need to have a love of the sciences, a never-ending desire to learn about the sciences, and the desire to share your experiences with others. Your rewards come from helping other researchers with their research."

He recommends "staying current using your existing resources in your field while expanding your affiliation and knowledge in related sciences. Knowledge and its continued pursuit is the key to this and any chemistry career."

Predictions

As for the future, Greg says, "Users of chemicals will always need technical advice on how to use them. As these needs become more diverse and sophisticated, they demand a greater degree of understanding of a specific research area as well as the ability to connect researchers to the most appropriate person to respond to their inquiry. Alternative ways of doing this will evolve that will encompass different levels of the business and vendor interaction. There also exists the opportunity for independent consultants to satisfy this type of technology-transfer need."

Ted Gast

Managing Director
Carl F. Gast Company

BA, Chemistry and German, Vanderbilt University, 1979
MS, Chemistry and Scientific Information, University of Cincinnati, 1986

Current Position

Ted Gast is the managing director of a family business that has been around for more than 70 years. The Carl F. Gast Company is a manufacturers' representative for industrial instruments and equipment, serving customers in parts of five states. The company specializes in selling technical products with non-self-evident uses to a wide range of industries, and as such the salespeople serve as local sales representatives for 10 non-competing companies that make monitors, alarms, and control systems; valves; and indicators usually installed in manufacturing plants as part of large control systems. The companies whose products they represent are called "principals."

Ted's areas of direct responsibility includes sales, customer correspondence, and communication with principals. He is also responsible for interviewing new hires, getting new product lines, watering the plants, and sweeping the warehouse. In this company of three full-time employees, there is a pretty good chance that if something needs to be done, whoever is there is going to be the one to do it.

A typical week begins with a trip to the office to read mail and e-mail, and to return phone calls. The rest of the morning is spent scheduling the remainder of the week. The sales funnel, which tracks the progress of all pending orders, is updated. Hot projects are reviewed. Appointments for in-person visits are arranged. Ted says, "A good salesman will make as many in-person calls with customers as he can, and 16 to 20 per week is a good number. That amounts to two each morning and two each afternoon. Cold calling is usually not productive, although it is better to make an additional call when you're already out of the office, instead of sitting at your desk all afternoon. One sales manager told me, 'Always make that extra call; what have you got to lose?'"

According to Ted, "A typical sales call goes like this: Meet the customer. Small talk. Discuss current and previous projects. Update database if any information has changed, especially new hires and changes to personnel. Ask what is the company doing. What projects are being worked on, and what problems are they having? Is there something we have that may solve their problem? Can we talk to the engineer who is designing the project and get him to specify our equipment in the plans? That means that when it comes time for the contractor to actually bid the job, he has to buy our products. These days most engineers are reluctant to specify one vendor exclusively, but it is always good to make sure you are at least on the list of possibilities.

"Once a job 'hits the street,' or goes out for bids, people are reluctant to change their specifications. First, it is more work for them, and second, people seem to come from nowhere to bid on the job, and many will promise anything but somehow are never to be found once things get going."

It is not necessary to leave the customer's office with an order. If the salesperson has done the job properly and built a good relationship, the order will come later. Long-term relationships are more important than short-term sales. Evaluate the call for follow-up, including the urgency. If necessary, prepare a quotation listing prices and availability for specific items that the customer is interested in purchasing. Since different customers may have negotiated different discounts, this may not be a trivial exercise.

Friday is another day spent mainly in the office on follow-up and updating customer files, so the decks are clear to start again on Monday.

Career Path

After college and a five-week tour of the Far East, Ted started work at the Seven Up Company. Ted initially chose industry over academia because of "family history." He says, "Gasts don't work in academia; they get a job and go into management." Ted's main responsibilities were running analytical tests, including doing a lot of titrations and water-quality testing. In addition, Ted syruped bottles for board of directors' tasting events, participated in taste tests (as did all other employees), and worked on expert panels. He worked on developing new flavors, including Citrus 7 and a failed cola project.

He also got involved in projects in packaging and engineering. He recalls, "That was my first exposure to marketing. The two biggest projects I had were the transition from bagged sugar to HFCS—high-fructose corn syrup—and the development of PET—polyethylene teraphthalte—as a packaging material—the now-familiar 2-liter plastic bottle. At that time, most of our product was still supplied to the consumer as a 12-ounce glass bottle."

Always one to take advantage of opportunities, and since Seven Up had 220 developers overseas, Ted used this as an opportunity to develop his language skills. Ted has studied 29 languages, including German, French, Spanish, Arabic, Uzbek, Sorb, and Mordvin.

At about this time, Seven Up was subject to one of the first corporate takeovers in the country (and one that has since been used in business schools as a case study of how *not* to do things). Since he would have been relocated anyway, Ted moved to Cincinnati, where his future wife was working.

There he worked at Fries & Fries, a division of Mallinckrodt, as an analytical chemist specializing in instrumentation and GC/MS. He found he had an aptitude for instrumentation and helped design and beta-test new equipment in collaboration with the manufacturers. Eventually, one of those manufacturers recognized his abilities, and Ted went to work there.

Ted started out at Tekmar as an environmental analyst, in the applications division. Ted also spent time doing marketing technical support, which involved helping salespeople close orders and answering customers' questions and solving their problems. There was some international and export work, and Ted's fluency with languages allowed him to get involved in that.

After some time, the president of Tekmar asked Ted to join the marketing department. Ted recalls, "This was hard for me since I was very comfortable in the lab. However, I think he saw something in me, and I think the other salesmen had told him how helpful I was in dealing with customers. Because of my lab experience I was good at closing sales. I thought that when the man in charge asks you to do something, it is a good idea to do it. So, I moved. One of my first assignments was to write an advertisement for one of our concentrators for the journal *Analytical Chemistry*. Up until that time, I had never really thought about anyone actually writing the copy for an advertisement, but I guess they really don't write themselves.

"The biggest change in moving to the marketing department was that I lost my laboratory. I still had access to it, but I had to work through my replacement rather than do things myself. If I had what I thought was a good idea, I still had to convince him that it was a good idea in order to get it done."

Being in marketing meant more travel, and Ted made several trips to trade shows in Germany—something he had never anticipated when he studied German. He says, "As my boss wisely told me, nothing you do is wasted."

In 1990, Tekmar had recently been acquired by Emerson Electric, and a sales position opened at the Carl F. Gast Company, which Ted's grandparents Carl and Lillian had founded. Even though he wasn't involved in the business anymore, Carl was delighted to have Ted join the family firm. Ted worked with his father and two other salespeople—the total staff of the company at the time was five people.

Ted recalls, "When I left to join the Gast Company, I thought, 'I've made a lot of money for other people; now I'm going to make some money for myself.'"

This change, from a large company to a small family firm, came with some adjustments. Ted points out, "When you work with family members, it is especially hard to leave the job at the office. Furthermore, working in a small office minimizes opportunities for socializing. Also, when you start out at the top of a small company there is not much room for advancement in the typical sense. There are the challenges of growing and maintaining the business and keeping up with trends and the changing economy and workforce, but no regular promotions or raises.

"Many people think that when you work for your own company you have no boss. It's really the exact opposite—I have 10 bosses, the principals. Some principals can be pretty demanding of my time. Being your own boss is a great challenge, but having control of your schedule makes it worth the trouble. Especially with children in school, it makes it all the easier to work around their needs. I also get to go home for lunch, although I'm not sure my wife sees that as an advantage."

There are also disadvantages—taking time off can be a problem. Any time spent not working will eventually and directly affect one's income. Ted notes, "This is not to say that there are no vacations; they just correspond with a drop in sales—and thus income—at some point in the future."

Ted feels that in order to have a successful career in technical sales, a technical background is essential, including chemistry, mathematics, physics, and engineering. Familiarity with laboratory practices and common industrial processes is a plus. Accounting can be beneficial but is not necessary. Ted observes, "You can take an engineer and make him a salesman, but you can't make a salesman into an engineer." People skills are a must to be a successful salesman—dealing with people individually. There is also a lot of travel required. Besides the sales call, inevitably something will go wrong with a piece of equipment, and one will have to make a site visit to investigate.

Ted says, "Responding to problems quickly is key. As soon as a problem has been brought to my attention, I visit the site to see exactly what the problem is and look for solutions. At this point I try not to assess blame, and make sure to always tell the truth and not cover anything up. I do have to remember that I am representing the manufacturer. I know the customer is probably under the gun to get things up and running again, so now is my chance to help. If it will take two weeks to fix the problem, I tell the customer that. He usually appreciates my honesty. If the problem seems obvious, I suggest something for the customer to try. If service is required, I offer to set it up through the factory. You may end up working with maintenance people, who often know more about a particular process than you do. However, you should always know more about your equipment than the customer. If something has been damaged, it may need to be sent back to the factory for repair. This is almost always cheaper, but it takes longer."

The salary for a salesman tends to run about the same as for those who stay at the bench, although without the opportunity for regular raises, and it may or may not come with company-paid benefits. Ted says, "There is a saying that sales makes money; everything else costs money. My job is to be responsible for the revenue that pays the bills and everyone else's salary."

Ted feels sales "has been a very rewarding career." He adds, "I was very fortunate to find something I was good at, enjoyed doing, and could make money at. It is very rare to find something that has all three components."

Advice

Ted advises, "My grandfather always said the key to his success was 'Work hard, make lots of calls, but don't fill out forms or surveys. Other people can do that for you.'

"Be aware of opportunities, and don't be afraid to take them."

Predictions

As for the future, Ted observes, "The sad thing is that with right-sizing and the changing economy, it is not often possible to maintain lasting relationships with people because they change jobs quickly.

"Instrumentation will remain important. There will always be a place for the measurement of temperature, flow, pressure, and level in all industrial processes. However, the industry is shrinking in the United States. How much of the industry will continue to move overseas remains to be seen."

Osman F. Güner

Executive Director, Cheminformatics and Rational Drug Design
Accelrys, Inc.

BS, Chemistry, Middle East Technical University, Ankara, Turkey, 1979
MS, Organic Chemistry, Middle East Technical University, Ankara, Turkey, 1981
PhD, Physical Organic Chemistry, Virginia Commonwealth University, 1986
Certificate, Executive Program for Scientists and Engineers, University of California, San Diego, June 2001

Current Position

As the executive director for Accelrys, a provider of computational science and informatics software, Osman has profit or loss responsibility for the company's "chemistry" business, one of the three business segments at Accelrys (the other two are biology and materials). He is primarily responsible for product planning and management, and marketing. Basically, his job is to make sure that the "right" products are being developed and released to the market. Osman says, "We want to make sure that the software solutions we develop will address a pressing need in the industry. In order to assure this, we need to have a profound understanding of not only the science and technology but also of how the scientists use these tools. We need to anticipate how the science and technology is evolving in order to prepare the software tools to be available when the market needs them."

A significant part of his job as product manager is to continually monitor and analyze the entire market and convert that information into prioritized marketing requirements for software solutions. The prioritization and reprioritization becomes part of day-to-day interaction with the software-development teams. Each product manager communicates his or her priorities to Osman, who synthesizes them into priorities for the chemistry business segment as a whole.

Another aspect of the job is scientific validation. Both case studies and benchmarking studies are performed in order to validate that the software tools they develop are accurate and effective. Most of these case studies result in publications in peer-reviewed journals, conference participation, and so on, which is why PhD-level scientists are recruited to manage and support the products.

Marketing includes many activities, from creating content for data sheets and brochures to setting product pricing and selecting packaging, developing promotions, and positioning and launching the product.

Most of Osman's time is spent interacting with the research and development team, managing day-to-day product-development and –release activities. Together with a team of product managers, the research and development team provides inbound as well as outbound marketing support.

Career Path

Osman's career in computational chemistry was a journey rather than a discreet beginning. During his master's work (which was primarily synthetic chemistry), he was exposed to some elementary computational tasks that intrigued him. Osman recalls, "I was thrilled with computers and the evolving computer technology. Having used a slide rule for most of my college work, I developed an interest in computers as they started to enter into our daily lives.

The first mainframe I used was an IBM 370. A computer that is less powerful than many PDAs nowadays, it was the state of the art back then. First I delved into programming, but I soon realized I enjoyed using the software applications more than programming. I liken it to being a good driver of a car, instead of a mechanic who deals with the engine. You can be a good driver without necessarily truly understanding the details and theory of the internal combustion engine."

When it was time to plan his doctoral projects, Osman made sure that he had both synthetic chemistry and computational chemistry tasks covered in the proposed research. He ended up having two supervisors, one supervising the wet chemistry work (organic chemist) and another one supervising the computational work (physical chemist). He earned his PhD in physical organic chemistry.

After graduate school, Osman moved to the University of Alabama at Birmingham and performed strictly computational chemistry research. This work was mainly doing quantum mechanical calculations in search of new materials to be used as rocket propellants and was sponsored by the U.S. Air Force.

In obtaining this position, Osman had successfully transitioned from pure synthetic chemistry to pure computational chemistry. He notes, "The career path that I followed is not typical, but in the early days of computer and information technology, such career paths did not exist. For someone who wants to move into the computational chemistry career today, there are numerous paths available. Some universities even have specific curriculum in computational chemistry or cheminformatics today."

Toward the end of his postdoctoral career, Osman had several offers to continue in the field of material science and one opportunity to move into life sciences, where he would be involved with the design of bioactive compounds rather than materials.

One night, while struggling with this decision, he had a dream in which he was a very old man, chatting with his young grandchild. He recounts, "She was asking me what I had accomplished in my life. I was either going to reply with 'helping people design better explosives' or 'helping people design better drugs.' The latter sounded better . . . and in the morning when I woke up, I had made my decision to move into life sciences. I have nothing against material science at all. In fact, I feel it is very attractive nowadays with the emerging opportunities in nanotechnology. But during those days, the only research I was working on was on high-energy, high-density materials, and moving into life sciences looked like a good idea. I have never regretted the decision."

Osman accepted the position of senior application scientist at MDL, a software company, where he worked for seven years. There, he provided primarily technical pre-sales support. He was called in by the salespeople to assist with the technical sales, to demonstrate the product and respond to technical or scientific questions. He also participated in many of the "outbound" marketing activities such as tradeshows, seminar series, conferences, and so on. In most software companies, application scientists are expected to also publish scientific papers and make conference presentations.

Eventually, Osman was promoted to product manager for ISIS/Host, the server foundation for a whole family of scientific-information products.

In the mid 1990s, MDL made a strategic decision to decrease investment in the three-dimensional (3-D) information-management area. Since he had been specifically recruited by MDL to promote the 3-D area, Osman felt this would be a good time to seek a position elsewhere.

He moved to Molecular Simulations, Inc., as a senior product manager for their small-molecule modeling products. Over the years, as the products grew and evolved, Osman's responsibilities increased. In 2001, MSI merged with four other companies to become Accelrys. After nine years there, Osman became executive director of cheminformatics and rational drug design, responsible for the company's entire chemistry-product portfolio.

Osman loves that his position lets him visualize concepts and be creative. He says, "The most intriguing successes of computational chemistry are reflected in computer-aided design (CADD) of new drugs in the pharmaceutical industry. Today, CADD has become an integral part of pharmaceutical research. The ability to contribute to the design of new and better drugs gives me a fulfillment that makes the effort well worth it."

Osman interacts with many scientists, managers, and directors in the pharmaceutical industry. Most of the interaction with colleagues comes during conferences, meetings, workshops, a variety of professional society activities, and visits to customer sites. Within the company, he interacts with the scientists and engineers involved in the development of software solutions, and with executive management interested in the state of the business and what, if any, changes are needed to make it better.

To work in this field, a very strong scientific background is necessary. Unlike many other software industries (like business, educational, games, management software), the computational-chemistry software market is very technical, and the end users of the products are very sophisticated. This means that the individuals designing, developing, and marketing the software must be equally sophisticated and must possess a combination of superior scientific knowledge (demonstrated by continued activity in the field through publications, etc.) and strong experience in software architecture and application development. Having advanced degrees in both chemistry and computer science is highly desirable. Second to that, a scientist with software development aspirations, or a computer engineer with science aspirations, would be most qualified.

Osman feels his background in physical-organic chemistry was instrumental in helping him adapt to a software-marketing role that targets chemical and pharmaceutical industries. Similarly, his experience in synthetic-organic chemistry has helped him work better with medicinal chemists.

Because computational chemists are typically part of multidisciplinary teams in the research labs, a good scientific understanding of multiple disciplines is very useful. Good communications skills are essential. Balancing in-depth knowledge in multiple disciplines with good social skills can be difficult, but Osman finds that part of the attraction.

Overall, Osman is happy with his career and the choices he has made. He says, "Every now and then, I wonder if I should have stayed in academia . . . then I discard the idea. I believe I have made the right choices at every intersection. Having firmly moved into the management and operational roles, I expect to continue on the same path to achieve more-senior management roles, eventually perhaps running my own company."

In fact, Osman recently started his own CADD consulting and training company, Turquoise Consulting.

Advice

Osman advises, "People who are seriously considering computational chemistry as a profession in the life sciences area should seek to improve their knowledge and understanding of biology."

Predictions

As for the future, Osman points out, "Computational chemistry is not a novelty anymore. It is an accepted aspect of research, especially in the life-sciences area. As it gets more and more into mainstream research, the demand for computational chemists will increase. The material-sciences field is now experiencing the challenges that were faced in life sciences 10–15 years ago, that is, the need to demonstrate a return on investments. However, with the emerging investment in nanotechnology, computational chemistry has an opportunity to become an essential aspect of materials research. I believe there are good opportunities in the materials-science area in the future.

"I think the whole area will continue to grow and expand. I believe we will soon start seeing degrees in computational chemistry awarded in some universities."

Dean Goddette

Director of Business Development
Rigaku Automation

BS, Biophysics, University of Connecticut, 1980
PhD, Medical Biochemistry, Washington University School of Medicine, 1985

Current Position

Dean Goddette is currently working as an independent consultant, specializing in business planning and market research for technology products in the life-sciences area. Typical projects range from small, short-term projects, such as conducting market analysis for a specific new product, to creating full business plans for start-up companies preparing to solicit funding. Dean says, "Market research and business planning requires understanding the problem in the marketplace, how it is currently handled, and the underlying technology. I have to understand how the potential new product or service will make things faster, easier, or better. This requires understanding the business issues and processes involved, how a new technology is rolled out into the marketplace, and the challenges involved in getting people to accept and adopt it."

Dean begins a typical day by checking e-mails and responding as needed. Then he starts making phone calls to find people and/or make arrangements to talk with them if a more substantial conversation is required. At least once a week he attends a networking meeting over lunch, to maintain and grow his list of contacts. The remainder of the day is spent researching, writing, editing, and making more phone calls.

Dean meets with clients in person as required, to update them on progress and obtain additional information needed to address their problems. He says, "I spend a great deal of time on the phone with different types of people, researching the issues involved in a particular project, until I thoroughly understand the unmet need in the marketplace. Understanding the need allows me to define who the customer is and thus identify the potential market for the product."

Career Path

During graduate school and in his postdoctoral position, Dean learned the skills involved in protein engineering, specifically, genetics, enzyme kinetics, and structure determination. He then went to work for Henkel KGA, in their protein-engineering lab. He used software from a variety of sources to research protein structure and function. He recalls, "This was a research position but with a fairly directed purpose. The goal was well defined but not the path to success. I was responsible for the protein structure and modeling work to determine the best mutations to make to achieve our goals."

Dean says, "Henkel did a great job of educating me, and all their researchers, about their business and how they looked for new business. Projects went forward or died based on their business justification, not the scientific interest. I originally got involved in the business analysis as a defensive move to protect the product I was working on." Also, one of the things Dean learned about himself was that he liked the idea of working with a team of people.

While doing protein engineering, Dean used a number of computational packages, mostly from Molecular Simulations. As it became clear that Henkel was going to wind down research at his location, he arranged a meeting with the president of MSI and expressed his interest in the company and how he thought that they could use a real user at their company to help better explain the value of their software. He was hired as a product manager at MSI and subsequently moved to Tripos for various computational chemistry products for both protein and small molecule drug design.

While at Dean was at Tripos, the Midwest sales representative moved to marketing and convinced Dean that it would be a good career move to take over the territory as a sales representative, so Dean spent a few years in the field selling one of those products. During that time he learned more about marketing than he had in all his previous years of marketing.

Dean says, "I also saw all sides of the conflicts inherent between sales, marketing, and development. In the best situations, these three groups work in concert to provide the right amount of information at the right time to help someone make a commitment to your company. Most companies, however, are far from the best situation. One difficult thing for a scientific person to understand is that not every communication requires the depth of a seminar. Sometimes a salesperson has one minute to identify an opportunity and enunciate their value proposition. Generally, it is difficult to imagine how a complex technical product can be summarized in a minute—often it can't, but that's all the time you have. Hence the need for marketing to be accurate but not necessarily precise. As scientists, we put a premium on being accurate and precise. Scientific papers are not limited by space or time—we expect that another scientist is going take the time to read the paper if he or she is interested. Even the titles can be quite lengthy to accurately reflect the content.

"But within a business context, your communication with potential customers is almost always limited by space and time. Hence the 'elevator speech': Tell me what you're offering and why I should care, and you have about 90 seconds. The conflicts often come from providing too much information too late. When a customer is calling for information, he or she usually needed it yesterday and 'It depends' is generally not a useful answer—even when it's true. "

After a while, Dean moved on to become head of marketing at a small biotech company in San Diego.

At about this time, his wife developed a life-threatening illness, and he quit his job to be home with her. He worked from home for a year at whatever consulting jobs he happened into. "That's what really took me into the consulting direction. My wife and I decided that we were not going to move anymore, and I decided to focus on consulting as a way to control my location and my job satisfaction. It was a great time to go into consulting. These days, companies are more willing to work in long-term relationships with consultants than they used to be. It gives them flexibility in managing their workforce and their expenses."

Dean does not compare his current salary to what he might have made had he followed another path. He says, "For me, it's more of a lifestyle issue than a salary issue. I have control of my time, and I work from home, and I really like that. The downside is that I don't have any guarantees as to where the next paycheck is coming from. But my experience in industry is that there is no guarantee there either; you just don't know it."

Dean observes, "My career path is not typical, but it came from following my interests. My personality is such that I have always worked at the interfaces between groups—between marketing and development, between sales and marketing—where my particular skill set of translating scientific jargon into benefits and advantages for a specific customer has come into play."

Dean interacts with a broad range of people, from scientists to CEOs to venture capitalists. Each group has a particular way of communicating, and they are each looking for a particular type of information. He notes, "With scientists, you can easily spend a great deal of time talking technical details, but with a CEO, I want to get right to the point. It is important for me to be able to adapt my communication type to accommodate other people's styles. Successful communication is incumbent on me, not them. It's important to be able to communicate with scientists but also to be able to translate the information so that a marketing or businessperson can understand it.

"One of the most important things from my chemistry background is having learned the difference between accuracy and precision as it relates to communication. Scientific communication is best when it is both accurate and precise. Marketing communication needs to be accurate but it does not necessarily need to be precise. By that I mean it must be true but not necessarily detailed."

Dean recently accepted a full-time position with one of his consulting clients. "Having worked with them for many months, I have a clearer idea of the people and the personalities involved." He prefers it to "taking a job 'cold' and moving into a new area."

Advice

Dean observes, "Chance favors the prepared mind. Be ready for whatever comes up, and don't be afraid to take chances.

"For people coming out of a technical background, cultivating the communication and selling skills needed to find clients or to convince people to move ahead with you is often a challenge. Most scientists think that selling is easy or that a product 'will sell itself.' I hear this all the time, and it displays only a complete lack of understanding of the selling process. Unfortunately, most people equate selling with used-car salespeople, and that is unfortunate. Selling is a science and a skill in and of itself. Take a course in selling skills at a community college or university extension program to get a basic understanding of what is involved.

"Keep in touch with your friends, acquaintances, colleagues, and so on throughout your professional career. Your network is your main asset. If I had it to do over, I would have spent more time maintaining and cultivating my network. I would not have let so many of my contacts lie fallow and fade away with time."

Predictions

"Over the last 10 years or more, the pharmaceutical industry has 'industrialized' and applied a number of high-throughput technologies to just about every aspect of research," notes Dean, adding, "There will continue to be opportunities for new technologies that contribute real value to improving the drug-discovery pipeline."

Patricia Hall Ward

President and Founder
Kingston-Ward Advisory Group
(previously Senior Account Manager, Spotfire, Inc.)

BS, Medical Technology, minor in chemistry, Shepherd University, Shepherdstown, West Virginia, 1976
Board Certification in Nuclear Medicine, ASCP, 1977
MBA, Hospital Administration, Southern Illinois University, 1978

Current Position

As a senior account manager, Trish handles the development and closure of sales and project management at her company's key clients. This requires a high level of expertise and experience in a number of different disciplines. She does everything from introducing the product to the customer planning accounts with the company, developing sales proposals, negotiating licenses, making technical presentations and demonstrations, closing multiyear contracts, training, providing technical support, and managing personnel and projects. In addition, she must maintain customers' goodwill and satisfaction with their account, and that part of the position is just as important as the more tangible pieces.

Trish says, "One of the main reasons I've succeeded in this field has to do with my technical background in both chemistry and biology, and the ability to understand exactly what a scientist is trying to accomplish. Therefore, I am better able to help them reach their goals using the products that I represent. Comprehension of the current trends in both chemistry and biology is critically important. This knowledge allows me to help effectively solve problems at the account and to partner strategically with the client on the company's behalf."

Trish is on site with her major clients one or two days a week. While there, she interacts with both the clients and the technical-support or services team to gain an understanding of how projects are moving toward the previously defined goals. She coordinates any new tasks with the technical teams and gets a sense of which new directions her product or services might help with. Once a product has been sold to a client, most of her time is spent expanding product sales and/or services. She notes, "This requires an almost 'ambassadorial' role at the account. It's important to keep everyone happy and moving in a positive direction for the benefit of both sides. I also keep a vision or set of goals in mind as I develop the business at the account."

Trish spends a great deal of time meeting with scientists and managers in various departments of the client company—everything from preclinical to purchasing and legal—so frequent conversations with these groups are the norm. She says, "Entertaining these disparate groups is another function of my job, as it builds a sense of camaraderie and trust. So, I have lunch or dinner with clients quite often and plan meetings between our mutual executive teams. In addition, all the information I learn has to get funneled back to the corresponding departments of my company, so I initiate teleconferences, e-mails, financial forecasts, and reports on a daily basis."

Trish lives near her accounts, not near her company's home office. She does travel to the home office to attend sales and product-strategy meetings. There she presents her region's forecast for business for the year and garners support for resources she will need to complete that business to meet (or exceed) her quota. Occasionally, she attends product training or conferences in the field. As a senior manager, she is also asked to prepare sales tools for representatives with less experience and to present these strategies at sales meetings. The marketing groups may ask her for feedback on new products or new trends in the market, so she spends time with these projects also.

Other typical projects include mentoring junior representatives, interacting with corresponding account managers from strategic-partner companies to effectively merge product lines for the benefit of the client, and drafting and executing account plans that are both tactical and strategic for the short- and long-term business with that company.

Trish says, "All activities from both the client and the company side go through me, so life can get very busy; the phone is always ringing!"

Career Path

When Trish graduated from college, she started a clinical rotation in nuclear medicine at two hospitals in Washington, DC. She enjoyed nuclear medicine as a field, because it was a new technology and very interesting. It also provided both a clinical in-vivo opportunity and an interesting clinical-chemistry lab opportunity. She loved the chemistry aspect of both areas and was mulling over whether to go to medical school or business school next.

She recalls, "One day as I was working on a new lab test for estrogen receptor sites, a sales representative for New England Nuclear told me that I 'belonged on the other side of the bench . . . in sales' and offered me an interview with his boss for a junior-level position working for their company but on-site at the National Institutes of Health. Since I had grown up in Bethesda, Maryland, and had always wanted to work at the NIH, I jumped at the opportunity. I apprenticed in sales at this huge international account and sold chemicals like there was no tomorrow!" Trish was not only selling chemicals but working at the bench alongside her customers, helping develop new assays. Not only did this allow her to really understand what her customers' needs were, they learned to trust her because she was one of them.

Trish learned early on that she didn't enjoy being in the lab by herself all day. She likes to be with other people working on a common goal or project and likes the freedom of managing her own schedule. Plus, she enjoys the diverse scientific endeavors that she encounters in sales. She says, "One day you can be working on the human genome project, and the next day you are working on a clinical adverse-events project with the FDA." She adds, "I doubled my salary the day that I joined a sales force, which was not a bad incentive."

After a few years, she moved to Water's Associate as their NIH–Washington area account manager, selling supplies, solvents, and equipment. Again, she worked with research groups who were her clients to optimize new assays for proteins, peptides, and chemicals.

In 1984, she moved to Pandex Laboratories and sold automated immunofluorescent assay systems and reagents. In 1986 she moved to Applied Biosystems, where she sold DNA, protein and peptide synthesizers, sequencers, HPLC equipment, and supplies for all of them. The automated-synthesis technology was just taking off at this time. She recalls, "It was quite a fast-paced time in my career; most days I sold $250,000 worth of systems!" Trish created lecture series around the scientists doing this work and helped get the technology recognized and widely accepted. She helped put together a conference, which became the Genomic Sequencing and Analysis Conference, now in its 17th year.

She believed in the technology and in the scientists who were using and developing it. "I guess the story there is that I've been known for helping key researchers achieve their goals by aligning with their vision and using the tools—chemicals, systems, and software—that I sell to meet these goals," Trish says. "I just don't give up on their dreams, and it's proven to be quite a differentiating aspect to my career. I 'partner' with the scientists and enjoy the formulation of assay and method development. It's where I use my skills as a chemist and biologist to understand their projects. They trust me and what I sell, because I communicate with them on their level."

In 1995, Trish left the equipment and supplies area and moved into software sales. She recalls, "I joined Molecular Simulations and sold computational software products in the mid-Atlantic region. It was a great learning experience, but I discovered that selling software was a high-relationship kind of sale. It helped greatly that I had a chemistry background, because I dealt with research and development chemists, mainly in pharmaceutical companies."

In 1997, she briefly joined Beckman Coulter to introduce their new capillary electrophoresis DNA sequencer to the market. This turned out to be more of a consulting job. She again worked with the top sequencing centers, placing these new systems alongside those of Applied Biosystems.

In 1998 she moved to Spotfire, at the time a brand-new start-up, to help get them into the pharmaceutical and life-science sector. Trish sold visual analytic software, which was originally a general product, but she "worked quite hard at developing new applications in science, and eventually closed global agreements for them with several pharmaceutical companies." She adds, "Their product is now one of the key products used by scientists in a number of life-science sectors and is considered a strategic tool for drug discovery."

As an account manager, she worked with a variety of people at the client sites, including midlevel scientists in management positions in R&D at pharmaceutical or biotech companies. Now she works with the upper-level executives in R&D, legal, and procurement for ongoing business deals. Within her own company, she works daily with a team of technical personnel—scientists or IT specialists—in the field or at the site. Because of the importance of her clients and the level of revenue they generate, she usually reports through a vice president of sales directly to the CEO and/or president of the company. She also interacts with marketing groups to tailor new products to the market's needs. It's also necessary to coordinate with customer-service reps in order to ensure that clients are receiving the best response.

Trish advises, "In order to be successful as a sales manager, you need to provide advice to clients who trust you. This is a quality that is inherent within your personality and really can't be 'taught.' Either you have it or you don't. Once you learn to merge the science of sales with this trait, you have the basis for a good career in this field. What makes that knowledge exceptional is the addition of a solid background in the sciences. This is what has made my career different from that of most sales managers."

Trish feels that her advanced knowledge of biochemistry and chemical processes has been indispensable. This technical information has allowed her to speak directly with scientists because she understands them and their needs. Also, she is at ease with scientific articles and papers and can grasp the content quickly. She relies on that ability when reading background information on a researcher's work before speaking with him or her, thereby achieving a level of rapport not possible for someone without a chemistry degree.

"My degree in science has also trained my mind to be ever curious, and that keeps me always searching the literature for the latest news and trends. It has helped me apply a logical set of steps to follow in any given situation, which is also very helpful," she says.

Trish has enjoyed the freedom to manage her own business domain over the years and appreciates the chance to work with top scientists in a number of fields. She notes, "I have had the satisfaction of truly being a part of some major scientific endeavors. Some of my clients have turned into lifelong friends, so that is wonderful. The financial rewards in terms of salaries, commissions, and stock have also been significant, and travel to fascinating cities around the world has been another a positive aspect of this career."

However, she does "not care for the necessary paperwork involved in managing business regions, and the continuous corporate meetings that always last too long!" Constant travel demands and all the uncertainties that go along with travel can be another daily challenge.

According to Trish, "The obvious challenge for anyone involved with sales is to continue to stay on top as the best manager. I have been able to consistently do this, but it does add stress to your life. Once you have proven yourself to be a major contributor to a company's revenue line, they tend to expect miracles every year. The higher level managers also start to resent the commission payments that you earn as well, and this necessitates starting all over with another company. Sad but true for all the top representatives in our field. The other challenge is to remain well educated in the science and in tune with changing trends in the market."

Overall, Trish feels that her career has been quite satisfying to date. But she says, "I might have considered joining a venture capital firm years ago. I would have enjoyed the business development aspect of that position, and it would have used all my scientific and business-management skills. It would be rewarding to help a number of start-ups succeed, rather than one at a time as I have done."

For the future, Trish has the option of remaining in a field managerial role or moving into an executive role in a corporation. Doing market analysis for capital-investment firms is also a possibility, as is moving into the larger industry sector of general informatics products and services. She could also consult for companies wishing to be successful at major accounts or to enter specific market niches. In fact she has recently started her own company doing just that.

In 2004, she decided to move out on her own and started the Kingston-Ward Advisory Group, consulting within the life sciences and federal markets on sales and management of sales people. Trish sees herself as "semiretired at this point, which is great." She adds, "Fortunately, my career has been so financially successful, based on my sales, that this is possible."

Trish has always preferred managing and coaching new sales representatives and so has remained in the sales field instead of taking the more traditional path and moving into corporate management. Trish says, "I just really enjoy working directly with the scientists and being rewarded for my success in selling excellent products to them. I earn much more than my internal colleagues, and I don't like to write reports and attend endless meetings, so the field is the place for me to perform best. Also, I believe that has been the best way to pass along my knowledge and mentor younger representatives and is the more rewarding path for me."

Advice

Trish advises, "I would highly recommend this career to anyone who wants to contribute to science without being in a lab every day. The niche of life-science informatics would be an especially good area to consider. It requires a blending of several disciplines that are of high value to companies and clients. The technology is always changing, which makes it fast-paced.

"While attending tradeshows, make sure to spend some time in the exhibit hall with the vendors to ask them about possible openings in their company. Develop a network of successful professionals and stay in touch with them.

"If you are inclined toward management, I would advise you to aggressively seek a managerial role as quickly as possible and to continue to move up through the executive ranks. Always keep an eye on any company that is shaking up the market and be prepared to join them to gain experience. Try all areas of sales, marketing, and business development to discover your particular expertise.

"I do think that it's important for people to keep moving along every five years or so, to keep fresh, and to move up the ladder with each job change. There's so much to learn, and I'm glad that I tried bench chemistry and in-vivo biochemistry; selling equipment, reagents, and software; and managing a business. Drug discovery, government laboratory work, chemical sales . . . all require different skill sets. Having a diverse skill set in sales makes you more valuable to another company, so it's easy to move around. A lot of women in the industry leave to do consulting, mainly so they don't have to deal with the company politics.

"Finally, take care of your clients and treat them as cherished friends. They will be loyal to you in return over the years. Go beyond the ordinary in this regard."

Predictions

Trish sees increasing demand for professionals in this field, pointing out, "With the flood of knowledge resulting from the last century's discoveries, there will always be a critical need for people who can decipher scientific information. Informatics will be in great demand, and the chemists and biologists who understand how to sell these tools will be in demand as well. Twenty years ago, there were no bioinformatic or cheminformatic scientists, so the field is rapidly changing.

"I still see the need for experienced managers, however, both at the start-up companies and at the older established ones that now need to reinvent themselves to adapt to the market. As there are fewer people with both science and business backgrounds, some of these firms have to choose among only business-school graduates, but nothing can replace the value of a science degree here."

ADDITIONAL RESOURCES

American Chemistry Council (www.americanchemistry.com) represents the leading companies engaged in the business of chemistry

ExecuNet (www.execunet.com) is a posts information about executive jobs, career issues, and executive-recruiting solutions

ISA (formerly the Instrument Society of America) (www.isa.org) is a nonprofit organization for automation professionals

Forbes/Wolfe (www.forbeswolfe.com) is an insider's blog on the science, markets, and undiscovered trends of nanotech

LinkedIn.com (www.linkedin.com) is an online networking system
Medzilla (www.medzilla.com) is a job board for biotech, pharmaceutical, health care, and science professionals
MIT Enterprise Forum (www.mitforumcambridge.org) is a technology-entrepreneur networking group
Nanotechnology Business Alliance (www.nanobusiness.org) creates a collective voice for the emerging small-tech industry
Rice Alliance for Technology and Entrepreneurship (www.alliance.rice.edu) supports entrepreneurs and early-stage technology ventures in Houston and elsewhere in Texas through education, collaboration, and research
Small Times (www.smalltimes.com) posts news about MEMS, nanotechnology, and microsystems
Technical Sales Association (www.technicalsalesassociation.org) is a professional organization for technical account managers
Ten Myths about Careers in Chemical Sales (pubs.acs.org/chemjobs/jobseeker/articles/print/chemicalsales.html)
Top Bio Jobs (www.topbiojobs.com) is a biotech job board

5 Chemistry and Business Development

THE PREVIOUS CHAPTER focused on sales and marketing—getting products and services to the customers who need them and providing support after the sale. This chapter will focus on the activities that happen earlier in the product cycle—deciding which products should be developed and which ones should be updated or discontinued, and dealing with the bigger issues such as what markets the company will pursue and what alliances it will make. The deals made are business-to-business, not business-to-consumer. In many ways, business-development people are looking at the bigger picture and focusing on the company's long-term future.

* * *

BUSINESS development is a broad field that includes many career possibilities. Generally, people in business development are responsible for finding ways to bring new business into the company. They can do so by developing new products and services or by taking existing ones into new markets. Business development focuses on implementing a strategic business plan through equity financing; acquisition or divestiture of technologies, products, and companies; and establishment of strategic partnerships where appropriate.

One of the main activities of all business-development professionals is to identify, research, analyze, and bring to market new businesses and products. Careful market research and competitive intelligence are used to identify unmet needs in the markets currently served or to suggest improvements for existing products. Either route will increase a company's product offerings and bring in additional revenue. Market research involves looking at a company's potential customers and understanding their needs and behaviors, with the ultimate aim of determining which products they are likely to purchase. Competitive intelligence can take this a step further to include analysis of other companies and those companies' products, customers, and marketing strategies. In either case, market structure, government regulations, economic trends, technological advances, and all sorts of other factors must be considered and their potential effect on future sales predicted.

In addition to creating new products for existing markets, business-development professionals identify, evaluate, and pursue the strategic and financial prospects of new market opportunities. Not only new customers within a given market but entirely new market populations can sometimes be identified that can be served with an existing, or perhaps slightly modified, product. Continual market research and close contact with the target market segment are required to identify and react quickly and appropriately to changes. Different market segments may require different marketing strategies or different pricing strategies. These issues must be carefully researched before one ventures into uncharted waters.

Developing long-term business plans, goals, and strategies to achieve those goals is also part of business development. This means playing the "what if" game—analyzing the possible outcomes of various scenarios and how the company should react to them to maximize benefits.

A major part of developing new business is establishing scientific and strategic partnerships, joint ventures, and alliances with other companies. Each side must have something of value to offer and something to gain. Once a potential partner with complementary business interests is identified and both companies' management agrees to work together, a deal must be structured so that each side gives something, each side gets something, and both sides win. The terms of the deal may be tentatively agreed upon before the companies enter into a substantial business arrangement, and then both companies will perform what is known as "due diligence." Due diligence is something like a background check on the other party to verify that all the information they provided is accurate and to identify any potential risks so that an informed decision can be made about the potential benefits and risks of the joint venture.

Ed Hodgkin, the president of BrainCells, Inc., says, "In biotech, business development is a licensing and alliances job. In other companies, business development may mean sales, but higher value sales—$1 million and up. In my opinion, business development is really thinking about strategy for what the business should be and ways to go out and get it. It's working at other end of the pipeline from product sales—we are trying to figure out what company should sell."

There is no one path to becoming a business-development person. People move into this field from all sorts of backgrounds and angles. In big pharmaceutical companies, most business-development people are former scientists who have chosen to leave the laboratory. Generally, they are development or discovery people who move over to the business side and work their way up. Since salaries on the business side are generally higher than those on the lab side, this is not an uncommon path. Most training occurs on the job, although business or marketing experience, in addition to a science background, is often desirable.

Coming from sales, law, or business may be a more straightforward path, because those professions are more used to negotiations and deal making. However, without a scientific background they may not really understand what the company is doing. Business development is all about how value is created; those involved must really understand the issues on a deep level.

The most important personality trait needed is the ability to form relationships—this is crucial. People want to work with people they like and trust, and the bigger the deal, the more important this becomes. Being able to negotiate the big deals and enjoying the back and forth aspect is also important. No two deals are ever alike because each one is crafted to meet the particular market dynamics of a particular industry segment, or the specific needs of the inventor or other involved parties.

Although salaries and commissions for business-development professionals have increased markedly, they may be tied to specific deliverables, such as a key contract or alliance. The marketplace is increasingly competitive, and companies that hire people to go out and get business for them expect to get good results.

Technology Transfer

Another new field, closely related to developing new business, is technology transfer. This term is used to describe the process whereby knowledge, facilities, or capabilities developed under federal R&D funding are transferred to public or private institutions (in the most general case, it is technical knowledge that is transferred from one entity to another). In some cases, a university has patented or copyrighted an idea developed by one of their faculty members. The university then grants or sells a license to that technology to another organization, which develops it into a commercial product. In the best case, everyone wins—a new, profitable company is created, and the university receives revenue to support more research. This field has taken off since 1980, when the Baye–Dole act was passed, which allowed commercialization of federally funded research.

The actual transfer of technology and products can occur in a number of ways, including through joint research agreements, cooperative research and development agreements (CRADAs), through outright exclusive or non-exclusive licensing, or many other creative arrangements. While this sounds simple, in practice it can be quite complicated. First of all, not all research inventions or ideas are patentable. If an invention is determined to be patentable, it can take months, and $10,000 or more, to file the patent. Technology-transfer experts perform technical and commercial evaluations of new technologies to determine whether or not the potential income from licensing will be enough to cover the time and money invested in obtaining the patent. Commercialization experts will be called upon to find potential business partners quickly to minimize the expense of filing patents in multiple jurisdictions.

As research funds become harder to get, cooperative agreements between researchers and industry are becoming increasingly common. License fees, royalties on products developed, and even cash from equity investments all provide valuable funding for additional research. These agreements are a careful blend of business and science that balance meticulous scientific research with the excitement of new commercial development. Every agreement is different, and creativity in developing them is an asset.

An additional benefit of these new, close relationships between researchers and business is that many students are exposed to business practices and concerns earlier, so they are better prepared for what to expect when they move into the business world.

Profiles

Ed Hodgkin

President and Chief Business Officer
BrainCells, Inc.

MA, Chemistry, Oxford University, 1985
PhD, Computational Chemistry, Oxford University, 1987

Current Position

Ed Hodgkin is currently the president and chief business officer of BrainCells, Inc., a start-up drug-discovery and development company. The firm currently has almost 20 employees and recently received over $17 million from venture capitalists.

Ed's responsibilities fall into two main areas—finding a drug candidate that the company can in-license to take to the clinic, and to raising money to support the development of that drug.

Ed explains, "Our company is based on studying neurogenesis using neural stem cells. We test compounds to see if they'll stimulate stem cells to grow and differentiate. We have a chief science officer, and people to manage the science. My job is to manage the business, to build a business model around our scientific expertise. I'm trying to find a drug that's been in Phase I trials at another company —so a lot of the preliminary work has already been done—that we can in-license for a different—or the same—research area." By targeting compounds that have already undergone significant testing at other companies, Ed's company avoids the ADME and toxicity problems that plague many drug development projects, when compounds that appear promising in the lab turn out to have undesirable biological properties. Ed says, "Many big pharmaceutical companies have now realized that they have valuable assets in compounds that have failed in clinical development for reasons other than safety. Companies like BrainCells, with a screening platform, provide an opportunity to reposition such assets in a new clinical indication."

Ed spends a significant amount of his time talking to biotech companies in the United States and Europe, many of whom have more projects in development than they can handle. Many of those companies want to outsource some of their projects, and they don't have expertise in neurogenesis, so it can be a good match.

Japan, however, is different. Ed notes, "Many Japanese companies are primarily interested in marketing compounds in Asia and want a marketing partner to handle marketing in the rest of the world. Since we allow them to keep Japanese rights, they prefer us to the big pharmaceutical companies that want all worldwide rights for themselves." Ed uses a business-development consultant to introduce his American company to appropriate Japanese companies. He says, "Japan is even more 'it's all who you know' than the rest of the world. You really need a Japanese person on your side to do business effectively in Japan."

In addition to searching worldwide for drug candidates and partner companies, Ed is constantly raising money to keep the company going. His first duty when he joined BrainCells was to write a business plan detailing the exact steps the company would take and the deadlines it would meet, and providing a detailed budget for meeting those deadlines. Funds then had to be raised to meet that budget.

Ed identifies potential investors from among his professional contacts and prior investors, and through other executives in the company. He sends all potential investors an executive summary of the company's business plan, and if he's lucky he will get an appointment to do a 1 ½– to 2–hour presentation. Ed will travel to the potential investors' locations, usually San Francisco, New York, or Boston. The investors who have an interest will then conduct due diligence, reviewing the intellectual property of the company, making sure it is secure and the company has freedom to operate, making sure the company's patents are not invalidated by prior art or other patents, and much more. Venture capitalists seek opinions on the company's prospects from outside experts, conduct detailed evaluations of the professional backgrounds of top managers, and may even ask for references.

Ed is on the road about a third of the time, meeting with people and talking about how they might work together. The rest of his time is spent in the office, on the phone, or in meetings, talking to collaborators or investors. Once two companies have agreed on how to proceed, Ed works with the lawyers to draw up contracts and nondisclosure agreements.

Since BrainCells is small, they use about a dozen consultants across all areas. Ed finds, retains, and manages the outside talent, mainly ex–big pharmaceutical people who know people and can help find compounds. They come in or work on a per-diem basis. Ed says, "As we get bigger, people will come to us, but for right now we have to find them. As we grow, we will need external experts less, but for now we're only paying for the time and expertise that we need."

Career Path

After earning a PhD in computational chemistry, Ed moved from England to the United States. He wanted to work abroad and was able to obtain a postdoctoral position at the Washington University School of Medicine with Professor Garland Marshall, one of the leading experts and pioneers in the field of computational chemistry. While there, Ed developed and applied computational methods to study structural problems related to drug discovery and biophysics. Some of his projects included conformational studies of HIV protease inhibitors and the simulation of helix transitions in transmembrane peptides.

After three years, Ed's visa was about to expire, so he moved back to the United Kingdom. He says, "Moving back to the U.K. was a difficult decision as I loved living in the United States. But, I liked the idea of moving back to Oxford where I had spent my student days."

Ed wanted to work in a dynamic small company where he could affect the drug-discovery process. He obtained a position with British Biotech, which had about 100 employees when he joined in 1990, and more than 300 when he left four years later. He recalls, "It was an exhilarating experience, working with some of the brightest people in the industry. I worked on two projects that moved to clinical development, Matrix Metalloprotienase inhibitors and Platelet Aggregation Factor antagonists. In both cases, the lead compound failed in Phase III. Earlier, I had toyed with the idea of an academic career, but once I met the people at British Biotech, I knew industry was the place for me."

In 1994, a headhunter recruited Ed to Wyeth. The Wyeth's Central Nervous System site in the United Kingdom was looking for someone to run its computational chemistry unit, and someone suggested Ed. He thought it was time to gain some experience in a large pharmaceutical company, and the position was right, so he moved.

Six months after he joined, Wyeth merged with American Cyanamid, closed the site near London where Ed worked, and offered him a position in Princeton, New Jersey. Once back in the United States, Ed quickly moved up the ranks to associate director of structural biology, where he managed not only the computational chemistry group but also the x-ray crystallography and protein NMR groups.

Ed moved to Tripos in 1999 for an operational, not a business, position. Ed recalls, "Initially I was senior director of contract research for the Americas and Asia, running a group of project managers. We had two shared-risk projects, which I managed, and I went out to try to get more contracts. This turned into a business-development role, as I went out and sold collaborations and found I had a flair for it. Eventually, I left management and moved into business development full-time—it seemed like a natural thing to do. Later, I was also given responsibility for Tripos' marketing group."

Ed sometimes missed doing science but "not enough to go back to it," he notes. "Most successful scientists become managers anyway."

Throughout Ed's time at Tripos, he interacted with start-up companies, and at very early stage he was drawn to the idea of joining such a company. While at a conference, Ed ran into a former Wyeth colleague who knew the venture capital firm involved in starting BrainCells. Ed contacted the investors and discussed the opportunity. The investors put him in touch with the founders, whom he met in San Diego shortly afterward. The company had "a combination of great science, top-tier investors already committed to the company and a highly experienced founding CEO," Ed recalls. "It was an opportunity too good to pass up. They liked me because I had experience in both the science and business aspects of drug discovery."

Ed accepted the position as president of BrainCells and moved to San Diego. In his business-development role, he enjoys meeting and interacting with a wide variety of people including those in "science, finance, business development, venture capital, public relations, and lots of people calling to sell their services." Although setting up deals and collaboration is exciting, there is also a downside. Ed says, "In executive management, hard situations arise, and you have to deal with them fast, which may not be pleasant—layoffs, underperforming people. Confrontational situations in negotiations are always challenging, and saying 'no' is not fun.

"Having a science background is important, because you have to understand the industry and fundamentally how it works. What are the key issues in getting a drug to market? Why is there attrition of drugs through pipeline? How is value created? What is the business strategy? Which market should we go after? How does the industry make money? You really need to have worked in the industry and to understand how it works. It would be very hard to go into business development right out of school.

"People who come right out of business school can work on valuation of a deal, but they aren't fluent in the language of the industry. They have structured thinking about the problems of growing a business, but chemistry training gives you much the same thing—you can use a scientific approach to business. But if you apply analytical thinking, and some math, you can arrive at the same conclusions."

Business development has a much shorter time line than does scientific research. Ed observes, "Wins happen on a time scale of days or weeks, not years. It's very dynamic, energetic, do it now! You must be in constant contact with everyone, able to be on the phone or on a plane at any hour. You must have a BlackBerry® to be in the business-development world. If you can't send e-mail to their

BlackBerry®, you are out of touch. In this field, the cell phone is outdated technology. It's a great pace if you love excitement and need shorter term satisfaction."

The financial gains are also better. Ed notes talented professionals can expect a "bigger salary, bonuses if performance is good, and often a piece of the action and the opportunity to make more money."

Ed also enjoys the traveling itself. He says, "Sitting on a plane is relaxing, because no one can get to you. That's the one place cell phones can't reach—at least for now. However, I don't like being away from home and not seeing my wife and kids for long periods of time."

Balancing one's work life with family life is hard in this business. Being constantly on-call can be quite stressful. "You have a few key things you need to get done for the success of your business, and the weight of the world is on your shoulders," Ed observes.

For the future, he says, "I want to stay in the entrepreneurial area, start-up biotech is perfect for me. Eventually, I want to be the CEO of a biotech company—that is absolutely possible from where I am. I might become a venture capitalist—this is also possible if I make a success of this company."

Advice

To those seeking a career in business development, Ed advises, "Learn the industry, and learn it well. An MBA is not necessary. Also, learn to rely on the expertise of others."

Predictions

"I think that deals between biotech and pharmaceutical companies will become more important for sustaining the industry," Ed predicts, adding, "Biotech is all about innovation, and the pharmaceutical companies are good at marketing, so it's a natural fit."

Jennifer L. Miller

Executive Director, Corporate Development
Amphora Discovery Corp.

BS, Computer Science and Chemistry, San Diego State University, California, 1991
PhD, Pharmaceutical Chemistry, University of California, San Francisco, 1996

Current Position

As executive director of Amphora, Jennifer participates in all the company's business-development activities. This includes identifying, pursuing, and negotiating opportunities to license their drug leads to big pharmaceutical companies and/or biotech companies; creating corporate slide presentations and product packages; and representing the corporate development department on the internal leadership team.

Jennifer's days come in two flavors, depending on whether she is in the office or on the road. Because she lives in California but works for a company on the East Coast, her mornings in the office are always spent answering lots of e-mails from her colleagues at headquarters. Once she's triaged the new e-mails, Jennifer sets about working on her current deals, entering into strategy discussions about potential new deals and following up leads. This is achieved through a combination of e-mail, phone calls, and meetings. Her late afternoons (after the East Coast has gone home) are generally quiet, and that's when she can devote more time to larger tasks, such as constructing company slide presentations, product packages, and so on.

Jennifer is on the road about 40% of the time, mainly for customer meetings. She says, "When we go to meet with another company, it is really critical that we understand what their expectations are for our presentation. It is my job to assess, and set, the client's expectations so there are no surprises once we stand up to present our company, technology, and products. Sometimes we are only presenting an overview, but other times it's an in-depth discussion of a particular drug-discovery program. It is an important detail that can significantly impact the relationship you have with the client company.

"Days when we're actually on the road are quite different from office days, except that I still need to find time to keep up on e-mails. My job on these days is to get my team—myself and a few other scientists who are along for the visit—to the customer site on time and well prepared. Sometimes this means I arrange for a meeting beforehand to cover key messages and roles. If it's a morning meeting, then we've typically had that discussion the night before. I am responsible for knowing where we need to go, how we're getting there, whom we are seeing, and what we are presenting. Once at the site, I make sure we have a computer set up with the slide presentation ready to go. My job during the meeting is really fun—I introduce our company and who will be presenting. I then listen, watch the room for reactions, take note of questions, and try to evaluate whether or not there is a potential deal with this company. Sometimes the business-development discussions happen during the visit, sometimes they happen upon my return to the office. After my team leaves the site, I always hold a debriefing where we compare notes and assess how well we did and what we'd like to do better the next time."

Overall, Jennifer's days involve varied tasks. She writes lots of e-mails and makes many phone calls, both to follow up on ongoing discussions and to "cold-call" prospective customers. She prepares corporate slide presentations and product packages and organizes and conducts science and business-development presentations at customer sites. She negotiates term sheets, confidential-disclosure agreements, material-transfer agreements, and other contracts. She also performs market research for new market areas the company is considering, and works to build consensus internally for go and no-go decisions regarding potential deals.

Career Path

During her junior year in college, Jennifer's biochemistry professor asked her to join a National Science Foundation (NSF)–funded Research Experience for Undergraduates (NSF–REU) program. She had a great time and says of the program, "It really opened my eyes to a whole different career path." The following year, her computer science advisor asked her to participate in a summer NSF–REU program in supercomputing at the San Diego Supercomputing Center. She was able to research a very different subject matter and again really enjoyed it; the contacts she made enabled her to obtain a position for the summer after graduation. She recalls, "I worked for and with three computational chemists, and they helped me prepare for my graduate studies."

She attended graduate school at the University of California in San Francisco, obtaining her degree in pharmaceutical chemistry, using computational techniques to study important biological processes. Her postdoctorate was short by design, since Jennifer always knew she "would end up in industry." She recalls, "The academic thing never really spoke to me, so once I had the PhD, it was just a question of which company. I wanted to stay in the Bay area and have always been attracted to small companies, and start-ups in particular."

Jennifer chose to start her industrial career at CombiChem, working as the computational chemist on various drug-discovery programs. Eventually, she went back to her computer science roots and moved to the algorithm-development group for a couple of years. In that position, she developed software and started to take on more management responsibilities. She says, "In retrospect, there were signs in my CombiChem tenure that I would be happy in the world of corporate development. I enjoyed interacting with our corporate partners, participating in due-diligence presentations, and working on strategic and tactical planning."

After a few years, CombiChem was acquired by DuPont and moved its offices. Jennifer had to commute 1 ½ hours each way, so she started looking around at other start-up companies in the Bay area. She received three job offers and joined the company with the most aggressive time line for becoming a drug-discovery company: Signature BioScience. Her job was to build a computer-aided drug-design (CADD) group to support the drug-discovery efforts. She explains, "My job required technical leadership in addition to management responsibilities. Signature underwent a tremendous amount of change during the time I was there . . . from 'high-flying biotech start-up' to company running out of money and closing the doors. Not fun but very educational!"

While she was at Signature, Jennifer got to know her colleagues in the business-development department, which led to her setting up pilot studies between the biotech companies she worked for, and becoming one of the people the company sent to do due diligence when they evaluated other opportunities. She says, "This allowed me to see a bit more of what the business side entailed. For the first time, I considered moving from the technical side over to business. The exposure to business-development activates piqued my interest in learning more about how deals are structured and negotiated. I did not expect this, but once I realized how much I enjoyed it, I looked for other chances to get involved. Eventually, I decided I wanted to work on the business side of things, instead of going up the science ladder, and work my way up to become CEO of a biotech or nanotechnology company.

"However, when the doors closed at Signature, we did not get any severance pay, and I needed to find a job!" The timing was bad. This was during the post-bubble "biotech winter," and Bay area biotech companies were not hiring. Jennifer did a lot of networking that summer to try to find a computational chemistry position there. She eventually made her way back to Amphora, which she had almost joined after leaving DuPont.

This time, Amphora really needed someone to do CADD in addition to project management. They didn't have a full-time position but agreed to a "long-term" contract for part-time consulting work. Jennifer says, "This was great for me, because I had an income, and I really liked the people and technology at Amphora. I was able to pick up a few other clients as well, and things went pretty smoothly for the next 6–9 months."

Jennifer did a lot of soul-searching during this time. It became clear that she needed to make a big change. She realized she enjoyed taking on leadership positions and had always been successful in them. As she thought about which skills she had and which she lacked, Jennifer realized that she needed to learn more about the "business of biotech." The question was how to get a job on the business side.

She started talking to people, and recalls, "The overwhelming advice I got was to make a lateral move inside a company. This seemed to be the only way for a scientist to gain business credibility—other than starting a company, which I almost did. All the while, I was getting more and more involved with Amphora. I really became a fan of what the team there had built, and it was easy for me to see what value their technology would bring to drug discovery. This was important—I'd interviewed for some business-development jobs where I did not feel that way about the technology, and I know that I have to believe in what I'm selling!"

Jennifer ended up approaching Amphora's management with a proposal. They wanted to extend her part-time contract, and she agreed, on the condition that they let her volunteer time in their business-development department, thus allowing her to learn the ropes and make the lateral move. The plan got off to a slow start because the company is headquartered in Research Triangle Park, NC, and she worked in the satellite office in California, but her responsibilities grew, and she eventually joined the company full-time as the executive director of corporate development.

Jennifer sums her job up, saying,"I facilitate the progression of partnering and licensing discussions from beginning to end, with the goal of bringing revenue into the company. My scientific skills continue to serve me well every day. For one thing, I have a good relationship with the scientists in the company and can act as a liaison between them and our businesspeople. I'm also able to provide more-technical evaluation of external opportunities that come our way."

Jennifer is enjoying her new career. She notes, "I love the challenge of always learning something new. This position really stretches me and is an education about the business of biotechnology. The frequent travel can be a bit exhausting at times. You have to be available all the time, and you have to stay on top of your correspondence, or deals fall through the cracks."

Jennifer interacts with all types of people—from bench scientists to chief executive officers and chief operating officers, and, of course, attorneys. She observes, "People skills are absolutely critical to this job. You have to be able to form a relationship with anyone. Dressing well and carrying yourself with confidence are important as well. Presentation and writing skills are key. Knowledge of typical deal structures is helpful."

Jennifer finds that her chemistry background also comes into play when she is selling or evaluating opportunities related to compounds. Her science background in general helps her be a more disciplined thinker when considering what opportunities there may be with a given partner and how to get a deal done, but her experience in drug discovery helps her present and evaluate opportunities.

One of the biggest challenges in this field is "remembering that you work for the investors, whether they are public or private," says Jennifer. "This is different than working for the scientists and trying to change the world, and it can be a hard adjustment. It's made a little easier by the fact that the salaries are good—you definitely do not have to take a pay cut—but your dry-cleaning bills do go up!"

There were other advantages to switching from scientific research to corporate business development. Jennifer says, "Working in corporate development will allow me more job opportunities in the future, because there are many companies that need experienced dealmakers.

"This story is still being written as I've only been doing the job for a year, but I think it's been a success. I love the rush of 'doing the deal.' I enjoy what I do and feel like I am contributing, learning, and, most importantly, moving toward my goal of running a company. I have a growing 'deal sheet' that will help me when I eventually move on to a new position. I also have a whole new appreciation for the real business of a venture-backed company."

Advice

Jennifer advises, "Dress for the job. Look for opportunities to get involved in meetings with external companies. Learn about deal structures and company valuations. Much of this information is available on the Web. Volunteer to help your corporate development department and express your interest in learning more—this will help you make the lateral move. Go to local networking meetings. Build relationships with the business-development folks you work with and otherwise know. Take them to lunch; pick their brains about opportunities. If you are just starting out, consider working for one of the big consulting firms. Be prepared for rejection—it happens a lot on the job as well. Mostly, be clear about what you want and why you want it. Then tell people about your goals. You are more likely to get what you want if you ask for it."

Predictions

Looking to the future, Jennifer notes, "There is a lot of innovation in small companies, and the R&D productivity of large pharmaceutical companies is in decline. We seem to be moving toward a model where large pharmaceutical companies will in-license more and more of their development pipelines from small biotechnology companies. In such a scenario, both sides will need people who can help evaluate and negotiate the transactions."

David A. Karohl

Director of Business Development
Carbon Nanotechnologies, Inc. (CNI)

BS, Chemical Engineering, Massachusetts Institute of Technology, 1985
BS, German and Engineering, Massachusetts Institute of Technology, 1985
MS, Chemical Engineering Practice, Massachusetts Institute of Technology, 1986
MBA, Rice University, 2001

Current Position

David Karohl works for Carbon Nanotechnologies, Inc. (CNI), a small company that sells Buckytubes (tubular fullerenes), which have exceptionally high electrical and thermal conductivity, strength, stiffness, and toughness. As a director of business development, it is David's responsibility to do everything necessary to generate product sales, in both the near and long term. This task includes representing the company to potential customers, promoting products, educating potential users of their products, and much more. David explains, "CNI is a materials supplier, and our products are always incorporated into other materials that are then converted into final articles. In this way, all our partners and customers build on our technology, and we add value to them by helping them learn to use our products in the highest performance and lowest cost ways."

As a company representative, David is often called upon to be the face of the company. He answers inquiries that come into the company via phone and e-mail, provides initial technical support to potential customers to help them select the appropriate products and product grades for their needs, and discusses custom products. David also represents the company and its products through conference calls and visits to prospective customers, and to more general audiences by speaking at conferences, tradeshows, and exhibitions.

Part of David's responsibility is to help coordinate product promotion, through advertising, the corporate Web site, e-mail and print newsletters, mailings, and so on. He must make sure that all messages from CNI are consistent, accurate, and cost-effective, and that they reach the target audience.

Miscellaneous tasks pop up occasionally. David negotiates nondisclosure agreements, sales contracts, and development agreements. He explains, "Development agreements are mechanisms for partnering that spell out the inputs and outputs of a product development project. Because our materials are unique, development is required in order to learn how to process them so as to maximize performance and minimize costs. Hence, we partner with other companies to develop a new or improved product. The agreement spells out what each party will put into the project—cash, human resources, equipment, product samples, analytical work, existing intellectual property—and how the parties will share the results of the project—who gets which patents or trade secrets, how the partner will purchase CNI products for use in the commercial product, and so on."

David is also responsible for setting prices for his products and monitoring other companies to ensure that his prices and offerings are competitive. He notes, "Setting prices for a truly revolutionary family of products is quite a challenge. We are developing very different grades of Buckytubes, tailored to very different uses and hope that CNI can share in the value created. For example, one gram or less of a particular grade of Buckytubes will go into very large-format field-emission TVs that will sell for many thousands of dollars—see the Samsung article in the November 2004 edition of *Technology Revue*—while a very different product can be used to make tons of electrically conductive polymer blends for use in equipment housings, seals, and so forth. We do our best to work with partners to understand the economics and make sure that all parties benefit. Our products can be used to create a longer lasting coating on a part, which results in a premium price for the coating manufacturer, which they in turn share with CNI when they purchase the Buckytube product."

In a typical day, David will talk to development partners, negotiate terms of deals in progress, and call or call on new prospects. He updates other management members of his projects' status daily. He also analyzes sales data and customers' technical results and prepares presentations and product-marketing materials.

Career Path

David began his career in production, then moved to engineering, and finally moved to the business side. He notes, "This was not a typical path, as most technical people stay in production or research and development, but it is not completely uncommon either."

Upon graduation from college, David interviewed with many companies and chose to join Solvay, an international chemical and pharmaceutical group, which had a particular project that suited his technical, cultural, and lifestyle interests. After a few months in Houston, getting to know the U.S. organization, David was posted to the company's world headquarters in Brussels, Belgium, to learn how to make the bleach in Tide with Bleach®. This inorganic-peroxide technology was developed in Europe, and their customers, including Procter & Gamble, were in the process of developing it in the United States. David's assignment was to learn the production process and process engineering and develop any new technology necessary for implementing U.S. manufacture. His group oversaw production in seven or so European countries, and David spent time in most of them, learning how they did things and conducting specific trials using their facilities. He also spent one year in production management at the factory in Jemeppe, Belgium. David recalls, "I enjoyed the expatriate life, the travel, and the technology, and became fluent in French."

Upon his return to the United States, David was promoted to lead process engineer on the project team to transfer the technology and build and start up the first plant of its kind in the United States. Shortly before start-up of the plant, David was promoted to production manager of a related product and assumed production management responsibility for the new product as well.

After that, David was promoted to engineering manager, responsible for the company's projects, budgeting, distribution, and market-support equipment, and the plants' environmental systems.

David had always been interested in the commercial side of companies, taking economics and marketing classes while in college and graduate school. His managers in Belgium had noticed this and mentioned it in his earliest performance reviews. After about seven years, Solvay approached David with an opening in marketing and business development for its most technical products and markets (e.g., high-purity hydrogen peroxide for semiconductor manufacturing, and organic peroxides for disinfection), and he made the switch from the technical to the business side.

Finally, because of consolidation in the division, David was offered a position coordinating sales and production in Solvay America's biggest business, high-density polyethylene.

David left Solvay in 1999 to go to business school full-time to broaden and deepen his nontechnical skills.

During his time in Houston, David followed Rick Smalley's work in carbon Buckyballs and the developing technology. He recalls, "A constellation of networking contacts at Rice and in the Houston chemical and polymer industry led to the initial contacts with CNI," and David was able to join CNI very early, as it was hiring staff in its transition from its academic roots at Rice to a profit-focused start-up. David says, "The opportunity offered work with top-notch colleagues—(Fortune 500 CEOs and CFOs and a Nobel prize–winning chairman of the board), cutting-edge business and technology—(Buckytubes), and a small-company environment, where wearing multiple hats is not just an option but rather a necessity." It was just what he was looking for. He says, "Here, we are building an international business around 'the defining polymer of the 21st century.'"

Looking back, David says he "should have left a large, risk-averse company earlier." But, he adds, "Unfortunately, in the petro and chemical industries, these are more the norm than the exception. I did not realize how much more I would enjoy the more active, exciting environment of a small start-up company."

David chose to move into marketing because he wanted more variety in his career. He is able to work with different types of people (including research and development staff, management, senior management, and marketing managers) and with many different companies. He is also called on to conduct both technical and business analyses. Being at a small company also means David is responsible for a variety of tasks; fewer staff means everyone must be something of a generalist.

David enjoys focusing on the business of science. He says, "One of the challenges of this field is having the focus to turn great ideas into real products. The 'cool' science and endless possibilities can distract from the near-term need to make money."

The part he does not enjoy as much is the travel, especially since he has small children at home. The amount of travel required varies significantly from year to year and from season to season, as the company evolves and its needs change. David points out, "Initially, as we put business plans together, there was very little travel. Implementing the initial plans entailed 25% or so travel for several years, to visit potential partners, speak at conferences, and get the word out. More recently, more partners and customers are choosing to travel to our site, so my travel in 2005 is actually down from past years."

David's experience has been that salaries in business and marketing are much more variable than those in production, with a much bigger upside but much more near-term risk. Also, start-up companies tend to offer less cash compensation but may offer more in perks, such as vacation flexibility and stock options.

Part of David's success was serendipitous. He observes, "The businesses I joined turned out to be in much higher growth areas than either the companies or I expected. In nanotechnology, the true revolution is just beginning. Materials are always limiting, so expanded materials properties lead to better existing products and enable brand-new products. The technology is truly disruptive. By being in the right place at the right time, I had lots of opportunities."

Fortunately, David had the skills needed to take advantage of those opportunities. He says, "The most important skills are in communication—speaking, writing, and listening. Basic computer skills are important, as are teamwork and the ability to conduct thorough and accurate analyses. Chemical engineering is a very broad discipline, and it places a strong emphasis on problem-solving steps, starting with defining the problem. That training has proven very valuable. In addition, the breadth of technical knowledge I received, and the self-knowledge to know when to ask the real experts, are both valuable. It's important to know what you know, and let others do what they do best."

In the future, David sees himself moving into senior management or an executive position in a small- to medium-sized technical company. He would also like to develop a more international career scope.

Advice

David advises, "Get a broad technical education, including both physical and biological sciences. Work on your people skills, and get at least basic business training in accounting, marketing, and so on."

Predictions

David is optimistic about the future, saying, "The future of nanotechnology is getting smaller everyday—which means it's getting bigger!"

James H. Wikel

Chief Technology Officer
Coalesix, Inc.

BS, Chemistry, Marshall University, Huntington, West Virginia, 1969
MS, Chemistry, Marshall University, Huntington, West Virginia, 1971

Current Position

Jim Wikel describes his position at Coalesix as "chief technical support, chief trainer, and chief pusher of the envelope." His company is developing a software tool to help in the early drug-discovery processes of lead generation and lead optimization.

He says, "Working in a small company provides a lot of different challenges and activities. I spend much of my time traveling. I live in Indiana and our office is in Cambridge, Massachusetts, so I have a long commute to the office. I don't need to be there all the time, since I am able to do much of my work over the Internet. As our product matures and we talk with clients, I meet with them in person to discuss how we can work together." Jim spends about half of his time on the road. This includes attending scientific and trade meetings and visiting client sites. Most trips are for two or three days, but occasionally he will be gone for up to two weeks.

On the sales circuit, Jim travels with the vice president of sales, who handles the business side of the discussions and sets the itineraries and agendas. Jim does the technical presentation and answers technical questions. Their presentations are generally given to computational and medicinal scientists who "not only see the product currently as it is being offered but are able to look beyond and extrapolate and see applications to their current problems, or to new ones that they would like to solve."

Since the installation of the product involves information technology (IT) and cutting-edge computer science technology, Jim also speaks to the basic IT requirements and explains how to implement them in the clients' specific infrastructure.

When Jim is not traveling, he is constantly testing the product. He looks not only for software bugs but also tries push the software to perform better and tries to identify new ways to use the product. He explains, "I look for improvements and prototype new methods or functionality for future versions. Keeping current with pharmaceutical technologies and computer science technologies is a must." He receives lots of good feedback and suggestions from potential clients and current customers.

Jim is interested in taking advantage of the technology to help solve problems. He says, "Many technologies have come into discovery with lots of hype as the best thing ever, but after the hype is gone, many assume a realistic role as a tool in the drug-discovery process. Most recently, high-throughput screening and combinatorial screening followed the same path—big hype, huge expectations, disappointment, and finally assuming an important role as part of the overall process. When I talk with clients, I think they realize I have walked in their shoes and have faced similar challenges. I try to represent the technology in a realistic framework. I do not set unrealistic expectations or tell clients how to do their jobs. The technology is a tool to help. Together we must find the right problem and set the correct expectations. Discussions are along the lines of how we can help with what we have to offer. Problems are never identical, and even the application of a single tool is seldom identical. Customization and enhancements are critical elements that require a collaborative environment."

Jim finds that doing business with pharmaceutical companies is a long process. Typically, the process takes months to unfold and requires multiple contacts and site visits. The first step is to identify a primary contact, usually through personal networking. The primary contact is viewed as someone who is technology friendly, visionary, and in a position of influence within their organization. "We contact this person by phone or in person. The first contact is to introduce who we are and what our technology may offer. The next step is to establish a meeting either one on one, or with a small group of scientists to provide more background on the technology and a demo of the application. These sessions are followed by internal discussions among the client's scientists to determine how and where the technology may be applied. Our goal at this point is to get the application into the hands of the client in order for them to try it in their environment and on their projects to determine the actual value-adding capability of our technology. We like to give the client a two- to three-month version of the application for evaluation. The evaluation period leads into a discussion of licensing and an expanded user base. Throughout this process, we attempt to establish a rapport with the client to understand their needs and to build a base of trust.

"Our technology is built around the concept of being very open and flexible. Our approach is to be helpful without requiring the client to change their processes. We want to fit into their process and to add significant value without causing a disruption. It is humbling to see the plethora of ideas and possible ways our technology could be applied. In order to provide the level of customization required by our client, we must understand their challenges and concerns. This generally proceeds in a stepwise fashion that requires listening and offering an approximate solution or workaround using our application in its present form. We then discuss the ideal solution and how we might make that happen. Our staff then discusses the options and designs a plan of implementation, often a tiered approach over time. This allows us to adjust as the needs change. This is the nature of research. The key is to be adaptive and responsive."

Career Path

Jim's career path has been one of seizing opportunities and taking risks. He recalls, "It was not typical but was evolutionary, reflecting the introduction of new technology. At my mid-career point, I was able to witness and participate in the beginning of a continuing paradigm shift in the pharmaceutical-discovery business."

Jim has always had an interest in science and during high school focused on chemistry. He says, "I guess the time was right, growing up in the age of Sputnik, rocket fuels, and early polymers. I grew up in southern West Virginia, in the heart of Appalachian coal fields, where the only people who knew science were the local science teachers."

Jim didn't want to teach, so he went to college to major in chemistry and minor in business, intending to become a salesman. At the time, it was considered a strange combination.

The job market was tight when he completed his undergraduate work, and he had gotten hooked on the experimental side of chemistry while doing his senior research project, so he continued in a master's program and jumped enthusiastically into research. His research advisor left for another academic institution after his first year, so Jim had to decide whether to start over with another professor, costing a full year, or to continue working with his boss supervising him remotely. He chose the latter and completed his research on time having only limited contact with his advisor. Jim credits this experience with teaching him how to work independently.

Upon completion of his master's degree, Jim obtained a position with Eli Lilly & Company as an organic chemist in discovery-chemistry research. He started operating very independently, as he had been doing. Fortunately, his boss allowed him this freedom, which was definitely not the norm.

After a few years, Jim decided he wanted to become a senior scientist and have a research lab of his own. He recalls, "I was completely naive about the glass ceiling that existed for non-PhDs. I was never pressured to get my PhD, but research is very top heavy, with virtually only PhD's occupying the higher corporate levels. As a result, BS and MS scientists who want to advance within the corporation usually moved into other areas. I stayed in research, worked hard, was promoted, and broke the glass ceiling in 1982."

During this time, Jim developed an interest in trying to quantitate and understand the relationship between structural changes and changes in biological activity, and he gradually became aware of the potential power of computers to correlate this data. He says, "I began my slow migration from the real world of experimental chemistry to the make-believe world of chemistry in the computer. I started hacking at computer programs to expand their utility, and soon other people started using my software. My code was not pretty, but some of it remained useful for over a decade."

In 1989, Lilly created a computational chemistry department, and Jim was offered the opportunity to join and pursue his quantitative structure-activity relationships (QSAR) interests full-time. He recalls, "This was the most difficult career decision for me, as I had no formal training in computers. How was I going to compete with those trained in theoretical chemistry? I took the risk, and it worked. I was able to develop the skills to understand, convey, and translate among these disciplines the viewpoints and requirements of medicinal chemists for cheminformatic or computational tools."

Jim established the formal QSAR group and over time found himself in a third career at Lilly, as a group leader and manager. He explains, "The role of a good manager is to allow his reports to be successful. My style of management was 'hands off.' I defined a strategy and a destination and then left many of the details in the hands of others. We hired the best people and best minds for the job, so I expected them to find their way. I preferred to have lively discussions and did not expect everyone to agree. My challenge as a manager was to promote free and honest discussion." Jim continued to advance and was head of the structural and computational sciences group when he retired.

He was retired for only a few months then was approached by the CEO of Coalesix about working with the new company, which had been created partially from a project Jim had been working on during his last year at Lilly. Jim accepted the new challenge to continue his research in multiparameter optimization and to develop computational tools useful to the medicinal chemists.

Overall, Jim says, "The most difficult part of my career choice was also the best. I was able to get into computer technology early as it applied to chemistry. This was and is an exciting place to be because completely new vistas are opening. Computational sciences continue to make contributions to the drug-discovery process, though the emphasis has shifted from basic theoretical approaches to more applied technologies. In general, computer technology in chemistry is an uphill struggle. Medicinal chemists are experimentalists by training and are often reluctant to trust a computer as much as they trust their own intuition. Anyone that pursues a career in computational chemistry or cheminformatics will find it a scientific challenge and a challenge in human psychology. The reward is in the success of getting others to use the methods you developed to gain new insights or to bring a new level of understanding to the data in bringing those methods to a broader audience."

Jim interacts with a diverse audience that includes medicinal chemists, biologists, pharmacologists, statisticians, computer scientists, information technologists, and management at all levels. He notes, "It is important to understand the subject matter at a deep level so you can talk with and understand the viewpoints of all these diverse individuals. Everyone has to win in order to make the collaborative environment work well."

In order to do this work, a strong technology-oriented mindset with an eye toward the practical application of the technology is a requirement. Jim observes, "Inventing or developing new technologies without widespread reduction to practice is a useless technology. Good people skills are required, as is the ability to work with a diverse population—requiring the ability to listen, blend ideas from others, and extrapolate into new space.

"The first 18 years I spent as a medicinal chemist in the lab and in the drug discovery process were paramount. They provided the foundation of knowledge I needed. I have to be able to think like and relate to the scientists involved in the experiments. I have to know their pain. This extends to the computer science side as well. I am able to read and write some code, which gives me a better understanding of what is involved from that perspective. Along these lines, I can describe what the method should do and describe what I think is the computer logic of the algorithm to employ. This helps get the idea across to achieve the desired results.

"The biggest challenge for the person in this field is to realize that this is a culture-changing opportunity and as such there are mountains to be moved. The pharmaceutical industry is under tremendous economic pressures to change its business. At the heart of this is the discovery and development process. Promoting change in any environment is difficult and always faces resistance. The basic pharmaceutical discovery process, successful in the past, has remained relatively unchanged over several decades. In the process, vast quantities of data are generated and used to support local decisions on a per-project basis. It has been difficult to understand these data at a higher level to leverage the learning across multiple areas. Converting data points into information and subsequently into knowledge and learning is the ultimate objective, and a goal as yet unrealized."

Jim now sees himself nearing the end of his career, which provides some freedom and new joys. He observes, "I have the freedom to explore a variety of ideas and continue to learn. I often think back on how fortunate I was in my early career to be associated with people who enabled me, and now I get a lot of joy from enabling others to be successful and achieve their goals."

Advice

For those with an interest in this field, Jim advises, "Academic training in chemistry, biology, and computer science is a great combination. Some of the best young scientists I have hired have trained in synthetic chemistry and computational chemistry. These individuals understand the language of chemistry and the practical problems associated with synthesis. Alternatively, a reverse combination of computer science and chemistry or biology is also an excellent combination. These individuals are able to convert a scientific idea into a software reality."

Predictions

Numerous opportunities exist for individuals who can understand the needs of the medicinal chemist involved in the drug-discovery process and also understand the computer technology needed to solve those problems. "The most successful individuals will have expertise in both sciences—chemistry and computers. There is a need not only for the technical side to be able to blend these disciplines but also for those who are visionary and can see beyond today and into tomorrow.

"Computer technology will not be going away. Science and technology are coming together on many fronts. Chemistry robots are doing compound synthesis, analytical instruments and purification instruments are using computer-based controllers and data-storage arrays, the industry is headed toward the analysis of genetic data for personalized medicines, and screening data with valuable and extensive toxicology data are stored within large databases awaiting useful analysis."

The current need are many. First, we need an understanding of the science around these data—knowing which questions can be asked or might be asked of these data. Second, the technology to extract knowledge from these data and what is the best technology available today to answer these demands is necessary. Third, we need an awareness of the evolution of technology of the future in order to stay ahead of the demand curve—to be proactive and not reactive. "We can all recount stories in which current data was stored in a form that was state of the art only to be faced with the realization later, when we had an important query, that the data were no longer usable because of technology changes. Or perhaps even worse, we were unable to query the data because it was not stored in a format that would permit our specific question."

Sadiq Shah

Director, Western Illinois Entrepreneurship Center
Western Illinois University

BS, Chemistry, Peshawer University, Pakistan, 1970
MS, Chemistry, University of the Pacific, Stockton, California, 1977

MS, Physical Chemistry, Washington State University, Pullman, Washington, 1982
PhD, Chemistry, Washington University, St. Louis, 1986

Current Position

Sadiq Shah directs a number of business development centers, all under the umbrella of the Western Illinois Entrepreneurship Center, part of Western Illinois University (WIU). There are three entrepreneurship centers (ECs), located in Macomb, Galesburg, and Quincy, Illinois. The mission of each EC is to facilitate the growth of start-up companies, and to support existing businesses so they can move to the next platform of growth. The ECs focus on developing educational programs, conducting research, and providing outreach with confidential counseling. The EC in Macomb develops programs mainly for WIU students, faculty, and staff, and for the surrounding three-county region. These programs are then delivered through all three centers. The Galesburg and Quincy centers are primarily focused on outreach, and each EC has its own manager to run the program locally.

Sadiq also directs the Office of Technology Transfer (which is responsible for technology transfer from the university to the private sector), the Small Business Development Center (which serves nontechnology-based start-ups or existing small businesses in a 12–county region), the Center for the Application of Information Technologies (which develops online educational programs and custom online training programs for corporate clients and state government agencies), the Illinois Manufacturing Extension Center (which works with manufacturing businesses to support their productivity needs, ISO certification, and so on), and the Procurement Technical Assistance Center (which works with existing businesses to guide them through the process of doing business with the government and the Macomb Executive Studies Center).

For each of these centers, and the Macomb Executive Studies Center (which develops and deliver professional development programs for management in regional businesses), Sadiq's responsibilities include budget and finance, strategic planning and execution of strategic plans, and development and implementation of new marketing and public relations programs.

He is also responsible for seeking grants to support many of their activities, including marketing and public relations. The Illinois Manufacturing Extension Center runs entirely on grant money. The ECs and the Small Business Development Center get half of their money from grants and half from the university. The other centers are funded by the university. Very little of the funding comes from the users—only the Macomb Executive Studies Center and the Center for the Application of Information Technology charge for their services.

For Sadiq, every day is different. He meets (on or off campus) with many people, including other organizations' representatives, to build partnerships for new initiatives; with staff to direct projects; or with various internal and external clients. On any given day, he may talk to entrepreneurs, students, faculty members, administrators, business owners, bank officers, venture capitalists, political leaders, economic-development officers, and other government officials in state agencies. He also interacts with colleagues in professional organizations.

Sadiq conducts certain workshops personally, including all of the business-development, intellectual property protection, marketing, and financing workshops. For the other workshops, he coordinates with the business school to identify suitable faculty to deliver the workshops together as a team. Occasionally, he has business faculty conduct a workshop alone.

Sadiq also writes grant proposals, quarterly reports, and final reports for various funding agencies, and reviews the financial reports of each department each month to ensure they are on track. He has supervises 10 staff members at the EC network, and each of the other centers has a director and managers who manage their departments and report to him.

Career Path

Sadiq received his master's degree in inorganic chemistry from Peshawer University and then taught at an undergraduate college in Peshawer. He moved to the Pakistan Institute for Nuclear Science and Technology and then taught for a time at Peshawer University.

Sadiq was close to finishing his graduate degree, and had a postdoctoral position lined up, when he received a job offer from Petrolite. He decided to accept, so he scheduled his dissertation defense, and on the following Monday started at Petrolite as a senior chemist responsible for developing a new technology. He succeeded in developing a new technology platform that the company still uses.

After about five years, he got a call from a recruiter about a group-leader position at Calgon Vestal Laboratories, a subsidiary of Merck. Sadiq interviewed, accepted the position, and spent the next 11 years working in R&D in the product and technology development for hard surfaces. Eventually, his responsibilities expanded to include skin-care and wound-management products and infection-control products, and the process development and product-delivery technologies groups also reported to him.

After about a decade, Sadiq started thinking he would like to make an impact on a broader scale than was possible developing technology and products in industry.

Throughout his career, Sadiq had assessed his skill base every few years to determine what he was prepared for and what new opportunities there might be. He had also continually identified areas of interest and educated himself on his own time and also on the job, by taking on new initiatives. He says, "It is important not to wait for an opportunity but to seek or create opportunities by identifying the needs of the organization and making a case for the position you want. You should not wait for new positions to be created; you should create positions for yourself with a well-thought-out strategy that will effectively support the organizational mission. Obviously, you must demonstrate an ability to successfully complete various projects and establish credibility by accomplishments on various fronts."

Sadiq planned his transition carefully. He wanted to leverage the broad experience base he had acquired during his industrial career. His technical work had brought him into contact with a number of different areas, including intellectual property protection, business agreements, business and marketing strategies, competitive intelligence, R&D management (in product and technology development and in engineering and process-development groups), regulatory requirements, marketing, and manufacturing.

He identified technology transfer as a new field in academia and felt his technical and corporate experience could help an academic institution. He joined the Association of University Technology Managers, a professional organization in this field, at his own expense to determine if his background would be appropriate. He attended their national meeting and talked to a few people there. He responded to two open position notices at the meeting and talked with the hiring administrators, who encouraged him to apply for the open positions.

Western Illinois University invited him to interview for the position of founding director of their new Office of Technology Transfer. He was asked to give a public lecture on what the next three years of technology transfer would look like, and he presented his vision. Since he had started new departments from scratch and built them in industry, he was an ideal candidate to establish a new technology transfer office.

Sadiq observes, "Technology transfer is a new field with lots of interesting opportunities to use my industrial experience. In industry, I had to focus on a narrow area in order to develop technology and products based on the company's business focus. Moving into technology transfer offered me an opportunity to make an impact on a much broader scale across many disciplines."

Once he started work as the founding director, he researched how offices at other universities were set up and studied what worked and what did not. During his first two months on he job, he put together a three-year strategic plan for the office. He started some entrepreneurial programs on campus to encourage technology transfer then started the entrepreneurship center; over time, he took responsibility for other departments.

Sadiq enjoys conceiving ideas for new initiatives and building partnerships and alliances to develop and implement new programs. He says, "I am enjoying the challenge and the opportunities to make an impact for the university. Guiding people on business start-up issues, business-growth issues, marketing issues, manufacturing challenges, Small Business Innovation Research (SBIR) and Small Business Technology Transfer (STTR) grant programs and the grant-application process is very personally rewarding, and I put in long hours."

Sadiq says his chemistry background gave him "analytical thinking skills," and, he says, "The PhD in chemistry has also helped establish my credibility. During my industrial research career, I learned business sense, interpersonal skills, communication skills, negotiation skills, fostering alliances, intellectual property protection laws, business-contract terms, grant-writing skills, oral presentations, understanding financial reports, planning, delegation, and understanding the bridge between science and business. Perhaps most importantly, I developed an understanding of business issues and how to identify creative ways to resolve them. Because of my scientific training, I am quick to learn the science before determining the business strategy.

"However, now I do more business than science. In fact, I do much more than technology transfer, also getting involved in developing strategies for the university in the online educational arena and getting involved in regional economic development.

"One of the biggest challenges in this field is the pressure to generate revenue and results. Technology transfer and business development is a business, not research, so you need to learn to think in different terms. Luckily, salaries are generally better than if you had stayed in the lab, depending on exactly where you end up."

Sadiq says he has "a broad experience base," and so could go in several possible directions. "However," he says, "I think that the rest of my career will be in academia on the business side, and more specifically in administration, more in a strategic planning mode. You cannot predict the future; however, you can shape it today. I enjoy what I do—I am helping people succeed in business, making an impact for the regional economy, and helping the university seize some opportunities in developing new programs and growing departments. I am having fun now."

Advice

Sadiq cautions, "Develop business understanding—just receiving or earning an MBA is not enough. Understand why marketing departments, finance, sales manufacturing, and so on, think a certain way. Be creative in identifying opportunities, and determine a match with your skills, based on not only your degree but also your job experience. Learn to market yourself based on your accomplishments and experience. Stay actively involved in professional organizations to develop your skills, and establish your credibility by successfully completing projects that are not 'me too.'

"The key is to recognize opportunities and have a plan as to how you can make a difference for the organization. Your skills and past efforts play an important role in establishing your credibility."

Predictions

Sadiq notes, "Having an understanding of the science, and bridging it with business, better prepares people for jobs in industry or in an academic setting. There will always be a need for those who have the science background but also have the business experience. Technology transfer is a relatively new field, and it is here to stay."

ADDITIONAL RESOURCES

Air Force Technology Transfer Handbook (www.afrl.af.mil/techtran_handbk_index.asp)

Association of University Technology Managers (www.autm.net) is a nonprofit professional association for intellectual property managers and business executives

BioSpace (biospace.com) provides up-to-date industry news, stock quotes, company information, job postings, and industry resources for all publicly traded biotechnology and pharmaceutical companies

CCL (www.ccl.net) is an electronic mailing list discussing issues in computational chemistry

CHMINF-L listserver (https://listserv.indiana.edu/cgi-bin/wa-lub.exe) is an electronic mailing list for chemical-information professionals

International QSAR and Modelling Society (www.ndsu.edu/qsar_soc/) is an organization for scientists who investigate quantitative structure-activity relationships in medicinal, agricultural, or environmental chemistry by classical QSAR, multivariate statistical modelling, molecular modelling, and computer-aided drug design

Licensing Executives Society (www.les.org) is a professional society for those engaged in the transfer, use, development, manufacture, and marketing of intellectual property

Network Science Resource Center (www.netsci.org) lists pharmaceutical-industry resources and articles and free online courses on drug-discovery topics

Professional Science Masters (www.sciencemasters.com) is open to those with a bachelor's degree in the sciences; this collection of two-year programs includes training in emerging or interdisciplinary fields that are designed to increase career opportunities

Science and Technology Entrepreneurship Program at Case Western Reserve University (www.step.case.edu) is a postgraduate program for scientists to build on their background and learn how to start new high-tech businesses or launch new products and services within existing companies and offers a two-year program to earn a master of science in chemistry entrepreneurship degree

Technology Transfer FAQ (www.nal.usda.gov/ttic/faq/t2faq.htm)

TinyTechJobs (www.tinytechjobs.com) posts news, career information, and jobs in nanotechnology, microtechnology, biotechnology, and information technology

VentureWire (www.venturewire.com) posts news and networking and other information about private technology companies, VC firms, and the people that manage and finance them

6 Chemistry and Regulatory Affairs

TO PROTECT HUMAN health and safety, governments pass laws to regulate product safety. However, in some areas changes occur so rapidly, and the specifications can be so detailed, that codifying them into law can take too long to be effective. In such cases, governments often set up regulatory agencies, which are empowered to enforce regulations that detail the critical technical, operational, and legal requirements for working in particular areas such as pharmaceutical or chemical. In this way, the public safety is protected, and new technologies can be incorporated into the process as fast as science allows.

In the United States alone there are more than 24 regulatory agencies that monitor everything from food and drug products, to energy and the environment. In each agency, there are employees who create and enforce the regulations, and they have counterparts in industry who monitor and ensure compliance with all applicable regulations. Probably the most well-known example of a regulated industry is the pharmaceutical industry, which is regulated in the United States by the Food and Drug Administration (FDA). Potential new pharmaceutical products must pass many levels of testing in the lab, then in animals, and finally in humans, before all the data is submitted to the FDA, which approves or denies the product for use by consumers for a specific ailment. Both the FDA and the pharmaceutical companies employ scientists to prepare and review the relevant data and interpret the results.

This field is called regulatory affairs, a relatively young profession whose specific boundaries and responsibilities are still being defined. At the broadest level, regulatory-affairs professionals oversee the progression of products throughout their life cycle, ensuring that they meet or exceed all applicable standards and requirements and pass needed tests.

In practice, regulatory affairs (RA) professionals are responsible for ensuring that all regulatory requirements are met, which include the requirements of local, state, national and international governing agencies. They carry out a variety of tasks, including determining which tests to conduct, collecting and analyzing data, and communicating the results of data analysis to various stakeholders within the company, the relevant regulatory agency, and other interested parties. They must ensure that appropriate testing has been conducted properly and that the results are scientifically valid. In some cases, RA professionals monitor market forces and evolving scientific conventions to ensure they are understood and addressed by all participants and incorporated into existing protocols.

Regulatory affairs professionals play a crucial role in balancing consumer-safety concerns with scientific advances in health care and derive great personal satisfaction from using their scientific knowledge to advance the public good—safely.

Consumers expect the products that they purchase will produce the effects promised by the manufacturer. Many classes of products must undergo several levels of rigorous testing to move from research laboratory to the marketplace. In order to ensure that the products are safe and effective, governments create both laws and regulations to define exactly what a product can be called and how it must be manufactured. Because of stringent regulations, companies that make pharmaceutical products, medical devices, and related products are those most likely to employ people to monitor regulations in specific areas, prepare submissions to regulatory agencies, and guide a product or product line through premarket approval, manufacturing, labeling, and on through to advertising and post-release surveillance. These regulatory-affairs professionals are involved in the entire product cycle, from development to widespread commercial use.

Some regulations apply to all companies—for example, Environmental Protection Agency (EPA) and Occupational Safety and Health Administration (OSHA)—but the more scientific a company's product, the more likely it is to need full-time professionals in this area. Clinical-research organizations, medical-device manufacturers, and especially pharmaceutical companies employ people to ensure compliance with regulations.

The pharmaceutical industry is one of the most highly regulated. Traditional tasks for regulatory-affairs professionals in a pharmaceutical company include preparing, submitting, and monitoring submissions (which can run into thousands of pages) to regulatory agencies, tracking the progress of the application through the agency, addressing issues raised by that process, and then monitoring products and regulations for post-release changes. As it develops, the regulatory-affairs field is becoming more science- and issue-based and less rote and bureaucratic. Regulatory-affairs professionals are becoming more involved at earlier stages of development, frequently helping design the studies and clinical trials that will produce the required data instead of just summarizing data that others have collected. They are also involved at later stages of product development to aid in the design of advertising, packaging, and distribution methods—especially when marketing internationally, where products must comply with multiple sets of (sometimes conflicting) regulations.

On the other side of the regulatory fence, governmental or regulatory agencies hire chemists, physicians, statisticians, and so on to evaluate submissions for new products, monitor compliance for existing products and processes, and develop new and better methods for product testing and environmental monitoring.

Working in regulatory affairs requires a broad knowledge of science, regulations, law, and business. Collecting data, collating it, and evaluating the results from scientific tests are all part of the RA professional's job. RA professionals must develop legal instincts, to be able to read and interpret complex regulations. They track changes in legislation and procedures in every jurisdiction that affects their products. Once a product has been approved and marketed, manufacturing procedures must be monitored to maintain product integrity and minimize environmental impact, and labeling and packaging must safeguard quality without adding too much cost.

The RA professional provides guidance throughout the product development process and often drafts and submits required regulatory documents. Timeliness and efficiency are crucial—the regulatory process cannot hold up product-development timeline. Professionals in this field need to be able to look at the big picture, see what the risks and benefits are from each possible course of action, determine how the plan will fit into the overall process, and explain how each action will affect the bottom line.

Because the field is relatively new, there are few formal educational curricula or certification programs. Most RA professionals have a bachelor's degree, and according to the Regulatory Affairs Professional Society, more than half have an advanced degree in a scientific or technical field. Most people transition into this career after working in medicine, research, manufacturing, or some other aspect of product development. Practical experience in research and discovery provides an awareness and understanding of technical issues, and training in practical analysis and logical evaluation. Being able to conduct critical, objective reviews; listen and observe carefully; ask insightful, clarifying questions; and then synthesize all the information gathered into effective strategies and plans of action is what scientists are trained to do, and these same skills are used by regulatory-affairs professionals.

Certification and professional degree programs are beginning to become available. The Regulatory Affairs Certification, an examination-based certification that meets both U.S. and European Union regulatory requirements, was introduced in 1990. The exam is currently administered every November. Continuing education and professional development are required to retain certification, once earned. Graduate programs in this field are beginning to be established, but the vast majority of currently practicing professionals have degrees in other fields. As programs become more readily available, official certification and degrees in regulatory affairs will become more important to potential employers. For now, many companies take internal employees who have expressed interest in this field and train them to do the work.

Any chemist can obtain a significant amount of information about regulatory processes, laws, and policies simply by investigating and reading. Reviewing company policies and procedures and understanding the regulations that affect the business can be a great place to start. Express an interest in the process and offer to help prepare regulatory submissions to ensure that they are accurate, clear, and sufficiently detailed. Take advantage of seminars, short courses, and classes to learn as much as possible about the agencies that regulate the industry.

In addition to a science background and familiarity with relevant regulations, an understanding of business principles is also important. Each potential project must be evaluated to determine not only which regulations will apply but also if the regulatory burden is going to outweigh the potential benefits. As with intellectual property, there are multiple ways to evaluate and represent a particular situation, and deciding which way to proceed can sometimes make the difference between a project's success and its failure. In some cases, particular technologies or practices will be selected because they have already been approved by regulatory agencies for particular applications. For example, the United States Pharmacopeia outlines in explicit detail the scientific tests a substance must pass to be marketed as a particular drug. Knowing which technologies to use, and being able to steer the company and projects to the most efficient ones, can directly affect the company's bottom line. Conversely, not pointing out potential regulatory nightmares will hurt the company's bottom line.

Meticulous attention to detail is very important. Working in regulatory affairs requires one follow a methodical, step-by-step process—even one missed step can have far-reaching consequences. It requires diligence and efficiency to cover all steps as quickly as possible, and personal integrity to ensure quality is not sacrificed for speed. In some cases, one must juggle conflicting priorities and deadlines while maintaining consistency and quality.

Record-keeping skills are also essential. Just as the rest of the world has gone electronic, so have regulatory submissions. This adds another complicating factor—keeping a tight rein on electronic documents and their revision status to ensure integrity throughout preparation, review, submission, and archiving.

Obviously, excellent written and verbal communication skills are essential, as is the ability to work effectively under pressure and deadlines. Good negotiating skills are often valuable, especially when working with the regulatory agencies, appealing and negotiating negative decisions, or requesting additional data for a new or currently marketed product. The ability to assimilate complex information and deliver just the key points to other individuals (scientists, doctors, regulators, patients) in a manner that is appropriate to each group is also important. For example, one must be able to explain regulatory requirements to senior management and to explain corporate decisions to regulatory personnel.

Good interpersonal skills are essential, especially relationship building, leadership, and negotiation. Building relationships with key agencies and personnel before disagreements arise is instrumental in resolving them quickly. Leadership skills are necessary to set time lines, define milestones, and set priorities for the various filings, most often in cooperation with the product-development team. In many cases, RA professionals work to educate other company employees about compliance requirements and help foster a cooperative, not competitive, workplace attitude.

Regulatory affairs can be a controversial job—regulators can be viewed as holding back progress, and innovators as caring more about getting their product to market than about patient safety. It can also involve significant amounts of travel—for example, to inspect remote manufacturing sites or conduct face-to-face negotiations with regulatory agencies.

Professional development and continuing education are crucial to a successful career in regulatory affairs. Policies, procedures, and requirements change rapidly, and professionals in this field must always be aware of the current situation, in which direction it is likely to change, and when. In addition, they must keep up with corporate and industry trends, often in more than one jurisdiction. Constantly changing requirements for different counties, states, and countries must all be monitored to ensure compliance with all applicable regulations.

With its intricate, overlapping, and constantly changing maze of regulations, professionals in this field never have to worry about being bored. Many are personally fulfilled by helping bring new products to market or by keeping dangerous products off the market. Whether viewed as a battle of wits by opposing factions or as a cooperative endeavor to arrive at the best outcome for all, there will always be a need for people to write, enforce, and apply regulations.

Profiles

Thomas Layloff

Principal Program Associate for Pharmaceutical Quality
Management Sciences for Health

BA, Psychology, and BS, Chemistry, Washington University, 1958
MS, Organic Chemistry, Washington University, 1961
PhD, Analytical Chemistry, University of Kansas, 1964

Current Position

Tom Layloff currently works at Management Sciences for Health, a private, nonprofit international organization working to improve management and access to critical services in primary health care, child survival, maternal and child health, family planning, and reproductive health, especially in underdeveloped countries. One of Tom's primary assignments was a project called Strategies to Enhance Access to Medicines (SEAM), which was funded by the Bill and Melinda Gates Foundation. He has worked in both Tanzania and Ghana.

In Tanzania, the emphasis is on pharmaceutical quality, and Tom works with the Tanzania Food and Drug Authority. Together, they developed training programs and Standard Operating Procedures (SOPs) to guide inspectors through examination of products and documentation at all ports of entry, including airports, seaports, and borders. The inspectional program also covers importers, warehouses, wholesalers, dispensaries, hospital pharmacies, Part I retail shops (which are full-service pharmacies), and Part II retail shops (which market primarily over-the-counter products). The inspectional-tracking program is transitioning from a paper form–driven system to a PDA Pendragon-based system whose data can be downloaded into a computer database. Tom is also helping establish programs at the Muhimbili University College for Health Sciences School of Pharmacy to train inspectors in how to perform colorimetric identification by thin-layer chromatography (TLC), a visual-detection assay, and visual-disintegration tests. As the SEAM project comes to an end, Tom is starting work on the supply chain for the President's Emergency Plan for AIDS Relief, which is funded through U.S. Agency for International Development (USAID).

In Ghana, Tom's project involves assisting the management of the Catholic Hospital Pharmacies to review their existing production and pooled procurement systems and to estimate the costs of these programs. They are also working to identify options to make them both more cost effective and more consistent with modern production concepts.

Career Path

Tom has always had an interest in science, especially chemistry and biology. He changed majors frequently in college and graduated as a psychology major who just happened to obtain a BS in chemistry. He had especially enjoyed synthetic organic chemistry, and a professor steered him toward graduate school in chemistry.

Once in graduate school, Tom found research and organic synthesis interesting and fun. He worked in the labs 12 hours a day, usually seven days a week. His advisor warned him to cut down his lab research and study physical chemistry, but Tom "bombed the physical chemistry exam," which, he said ended his graduate career at Washington University. Tom did not know what to do, but another professor suggested he go to the University of Kansas and work with a prominent organic chemist. Tom started there, but after his first semester the professor he had hoped to work with moved to another institution. Eventually, he went into a lab performing electron spin resonance (ESR) studies of electrochemical intermediates in solution. Tom found it fascinating, switched to analytical chemistry and completed a PhD in ESR.

Tom recalls, "When I was completing my PhD there were many academic jobs for analytical chemists from our group. Our research professor doled out interviews to whomever he thought would fit in the various schools. One of my lab mates interviewed at St. Louis University and was not interested in the position. He asked me if I wanted to return to St. Louis. I decided that I did, so I interviewed and was hired. They agreed that I would teach general chemistry for two years, and after that only teach instrumental analysis and graduate-level courses."

That worked for two years, until the senior analytical chemist had a heart attack and Tom had to cover the quantitative-analysis course, which he disliked. To make it more bearable, Tom hired undergraduate lab instructors and worked with them over the summer until they could perform every required experiment perfectly and could answer any questions about the course. This unburdened Tom from "the day-to-day noise associated with large undergraduate classes."

About this time, the FDA commissioner decided that every FDA field lab should have a science advisor to help improve their scientific base. Since most of the field labs performed regulatory analysis, they wanted analytical chemists. Tom was asked to serve as the science advisor for the St. Louis District Laboratory. His consulting centered on methods of instrumental analysis and included presentations, informal visits with staff members, and occasionally attendance at meetings with other science advisors.

Shortly after he started as an advisor, the St. Louis laboratory's primary focus shifted from performing food analysis to conducting national drug-surveillance programs. As part of the change, Tom's position became part of the FDA Center for Drugs operations, working with the National Center for Drug Analysis. Tom recalls, "The change in focus placed significant demands on my time, since the chemists had minimal training in the types of organic-chemical analysis required for drugs." The transition went exceptionally well, and Tom enjoyed working on drug analysis. After a couple of years both the director and deputy director of the center retired within a few months of each other, and the FDA lab was looking for a new director.

After giving it considerable thought, Tom applied through the Civil Service Commission for the directorship and was selected to fill the position. He decided to resign his tenured professor position and accept the new position as director of the National Center for Drug Analysis. He recalls, "I elected to resign my academic position to force my intellectual focus on that position only—I mentally closed the option of an easy return to academia." The FDA laboratory had a very stable, dedicated staff, most of whom relished the challenges of dealing with the massive bioavailability problems with marketed Digoxin products, the generic-drug fraud issues, and numerous other pressing issues at that time. The laboratory became an FDA showcase for regulatory analysis, and scores of foreign visiting scientists came there to work with them for as long as a year, learning automated drug analysis and drug dissolution techniques.

There were numerous international collaborations as well. Tom oversaw a 10-year project assisting the National Organization of Drug Control and Research in Cairo with their regulatory issues and helping the Egyptian government improve drug manufacture in their government-owned facilities. An eight-year program involved working with the Ministries of Health and Commerce in Saudi Arabia to improve their regulatory-laboratory operations. This job entailed sending staff from his laboratory work in Saudi Arabia for up to a year. The National Center for Drug Analysis had numerous visiting scientists from the People's Republic of China, and Tom spent a month in Beijing lecturing on U.S. drug-control issues.

In 1999, Tom transferred to the headquarters in Rockville, Maryland, where he served as associate director for regulatory standards. There he worked on issues regarding harmonizing U.S. regulations with those of Europe and Japan. The FDA was originally involved in the development of public standards for chemical and pharmaceutical products, but eventually the standards programs were turned over to the United States Pharmacopeia (USP), and the FDA focused only on safety and efficacy issues.

The USP is a not-for-profit corporation responsible for establishing appropriate product quality–assessment standards for marketed drug products and also for establishing which materials can be used as references in performing those assessments. Basically, the USP specifies exactly which tests a substance must pass in order to be sold as a particular product. The FDA encouraged Tom to volunteer with the USP, and he was eventually elected to serve two five-year terms on the Committee of Revision and on the executive committee, which sets policy for the public-standards activities.

That volunteer work lead to a paying position when Tom retired from the FDA. He joined the United States Pharmacopeial Convention as vice-president and director of the pharmaceutical division. While at the USP, Tom was responsible for three major areas—establishing and revising the product standards published as the U.S. Pharmacopeia/National Formulary (USP/NF), testing proposed reference materials and establishing stability-testing programs for them, and investigating monograph-test complaints and new methods. He supervised about 150 people in the three departments under his jurisdiction.

Tom left the USP in June 2001. Since he had left St. Louis University on friendly terms, he was able to return there as an adjunct faculty member. His plan was to do some teaching and organize a research effort on silkworms and the thermodynamics of chick development. He also wanted to organize some weekend retreats on technology-society issues and planned to do some consulting as well.

About this time, another former USP employee asked Tom to join him at Management Sciences for Health (MSH) to help improve health systems in the developing world. Tom recalls, "The choice became whether to pursue my intellectual interests or to focus my time on helping others. I agreed to work with MSH on development and have been busy in this activity for the past four years." Tom was able to bring his experience in pharmaceutical regulation and law enforcement and his training in analytical methods to developing countries. All his background comes into play when helping build and improve regulatory systems in underdeveloped countries.

At MSH, Tom is a technical expert and therefore has no managerial responsibilities. Managers in various countries seek his assistance in establishing or improving drug-regulatory programs. He interacts with individuals ranging from ministers of health, to government-operations and pharmaceutical-control officials. He also interacts with his MSH colleagues, who tend to be idealistic and dedicated to trying to help the less fortunate. Tom has been the only person at MSH to deal with drug-regulation issues, though he is now working with several other employees to help them learn to address some of these issues.

Tom's Tanzania project is a long-term transition, which involves program planning, project management, and both personnel and institutional development. These tasks are generally performed in Tanzania and involve a third to half of Tom's time. The Ghana projects are less intense and involve about 10% of his time and periodic two-week trips. Other assignments are short-term, requiring several weeks of travel followed by a few weeks of report preparation and meetings.

Tom also works on USAID projects to improve pharmaceutical quality, a tuberculosis drug-quality project in Kazakhstan, HIV-AIDS drug-quality issues in Ethiopia, and drug-registration strategies for the Ministry of Health in Namibia. He has represented MSH at the HIV-AIDS Fixed Dose Combination Product Therapy conference in Botswana and at drug-registration harmonization efforts in the Southern Africa Development Community.

Tom is currently serving as the quality-assurance manager for the Supply Chain Management System supporting the President's Emergency Plan for AIDS Relief. This is a monstrous undertaking to supply HIV/AIDS drugs and test kits, and medical supplies, to patients in 15 developing countries.

While Tom enjoys the work, it does come with a price. He reflects, "Almost all the people I work with are pleased that I have come to help them and I find that very rewarding—to be able to fulfill a worthwhile mission and have everyone pleased with the effort. However, the inevitable heavy travel schedule has dominated almost all my activities and has left no time to pursue my intellectual activities at the university."

In the future, Tom thinks he "likely will become more deeply involved in assisting developing countries establish viable and sustainable regulatory functions to help protect their people from unscrupulous drug merchants of the world. It is a deeply rewarding career, and I'm having a great time."

As he has done all along, Tom will continue to take advantage of unexpected opportunities. Tom says he never consciously decided to change professions, he "just ended up doing something else."

Advice

Tom says, "Staying flexible and being in a continuous learning mode" is a challenge but is essential for career growth.

He adds, "If you're interested in international regulation, consider working in the Peace Corps after your degree to get international experience. Some time in FDA field operations and drug approvals and compliance, and a year or two with the World Health Organization, getting oriented to the scope of problems, would provide an excellent and well-rounded background. Before you know it you have a career in international public health!"

Predictions

"As the population and globalization grow there will be unlimited opportunities for growth and development in industry, regulation, and all public health areas," Tom says. "The pharmaceutical and biopharmaceutical product markets will continue to grow worldwide as will the globalization of the market. There will continue to be growing opportunities in both the industry and in regulatory agencies."

Lucinda F. Buhse

Director, Division of Pharmaceutical Analysis, Center for Drug Evaluation and Research
U.S. Food and Drug Administration

BA, Chemistry, Grinnell College, 1982
PhD, Physical Chemistry, University of California, Berkeley, 1986

Current Position

Cindy Buhse is the director of the Division of Pharmaceutical Analysis, which is part of the Center for Drug Evaluation and Research section of the Food and Drug Administration (FDA). She, along with her staff of 30, work to ensure the quality of drugs in the marketplace through analytical methods development, method validation, and drug analysis. Some of their work is "immediate response," for example, determining the quality of more than 300 samples of KI tablets (to be used by the government in case of a dirty bomb) in only four days over a Fourth of July weekend, and determining the quality of drugs purchased from foreign Internet sites. Her division also conducts longer term research projects, such as developing new methods and criteria for assessing the quality of various dosage forms, including inhalation products and transdermal products ("patches").

Cindy spends most of her time mentoring her staff and setting policy and directions for future research. She also edits her staff's reports and manuscripts, advises team leaders and chemists on how best to proceed with their projects, and organizes collaborations with outside companies and organizations and monitors their progress.

In order to ensure that her division is progressing in line with agency policies, Cindy often has teleconferences about agency issues with her counterparts in Maryland and spends about one week per month in Maryland meeting with other agency officials to discuss issues and set policies.

She says, "We try to make everything we do relevant. Some of our research ends up helping when writing guidance—for example, there is now a nasal inhalation–product guidance based on some of our work, and work we did on topical products is changing the definitions of creams, ointments, lotions, and so on. Our study on the quality of drugs purchased over the Internet ended up in front of Congress. We help with data for citizens' petitions—we just finished looking at lead in pharmaceuticals for one of those. We are pretty free to do what we want with research into analytical methods for pharmaceuticals, but we try to keep it focused on relevant work. Otherwise it would be harder to justify our existence! We periodically talk to the different offices about what they think is missing in terms of analytical chemistry of pharmaceuticals–such as methods to determine bioequivalence, and so on—and then we try to see if we can develop something."

Cindy communicates with people at all levels from the janitor to the center's director, academicians, industry partners, and so on. She says, "It's mostly scientists like myself—chemists, pharmacologists, doctors, toxicologists—but I do need to know how to relate to office and janitorial staff and get them to do what needs to be done."

Career Path

Besides chemistry, Cindy learned some valuable skills in graduate school, including how to deal with people. She recalls, "We had quite a few contretemps in grad school, plus my graduate advisor left in the middle of my tenure, so a couple of us had to finish on our own. This really taught me how to get things done without support. It also taught me that you have to look out for yourself." She learned how to solve problems and got a good grounding in technical writing.

Even while in graduate school, Cindy knew she did not want to go into academics. She says, "I was not wild about teaching, and there was not much teamwork at Berkeley—a lot of infighting within our group and between professors —which kind of turned me off of academia." She had an on-campus interview with Rohm and Haas and, upon graduation, accepted their offer and moved into industry.

Cindy got started in supervision at Rohm and Haas almost immediately, when she was assigned a couple of technicians and bachelor chemists to work with her. She found she was able to get them to do good work, and people seemed to enjoy working for her. Cindy soon realized that she was a "facilitator–people person," but it was her peers with technical successes who were getting promoted.

At about this time, she relocated to follow her husband's career. She decided to take advantage of the change to look for a supervisory position, not a technical one, at a new company. It turned out that Cindy's grandmother knew the CEO of Sigma-Aldrich, one of the biggest employers of chemists in her new location. She says, "It took a lot of effort to call him out of the blue and ask about opportunities. I tried to keep it general and did not come right out and ask him for a job, but the conversation did get me an interview."

Cindy started at Sigma-Aldrich as a validation engineer in the diagnostics department, with the understanding if she did well for a few months, she would become production supervisor. She did well and was promoted into her first chance to supervise a larger group of six or seven chemists—both BS and MS. While there, she learned about Good Manufacturing Practices (GMP), a set of checks, balances, and controls to ensure safe, effective final products.

After a short period, she was promoted to production manager, which meant supervising supervisors. She recalls, "It was very interesting. I had to learn to back my supervisors even if they weren't particularly good at supervising!"

Cindy also learned that production ran at a much faster pace than research, which took some adjusting. She says, "Production is all about back orders and end-of-quarter push. Priorities change every day—or hour. It really teaches you to be flexible—something you need less of in R&D. You learn how to solve problems quickly and make decisions quickly. Sometimes you need to just pick a direction and move! Production really helped me see clearly what is and isn't important, because you don't have time or resources to do anything that won't improve the bottom line. I also learned a lot about business and finance . . . all those terms like ROI, accounts payable, and so on."

From there, Cindy moved on to become a validation manager, which involved supervising engineers, a much more technically challenging position in which she learned about equipment and HVAC systems. When the validation department was eliminated, Cindy became a quality-control manager in the fine chemicals department—she was back to using her GMP expertise and supervising bachelor and PhD chemists. "There is a little difference in supervising PhD chemists . . . they need less direction in some ways—technical, but more in others—being reminded that the focus is on the customer's needs, and they may have to leave a technically interesting question behind."

After a while, the senior management changed at Sigma, and Cindy began looking for something new. She saw an advertisement in *Chemical and Engineering News* for a job at the FDA and applied. The analytical experience from her last position, and her years of supervisory experience, were crucial to her being offered the FDA position.

Cindy believes that "management is similar whether you are supervising hourly workers, chemists, or engineers." She adds, "It is a matter of motivating people to do their work in a timely fashion, prioritizing, and ensuring high-quality work. I love to organize and arrange things, but sometimes people do drive me nuts. Generally they don't like change, and many don't put the job first—or even second, which is frustrating. You often don't have the people you need to get a job done, but you don't have the money to hire new people, so you need to get people to adapt, change, and so on.

"One of the biggest challenges I see is flattening organizations. Middle management is often squeezed out during cost containment and such. Even in government they talk about reducing the number of layers of management. We constantly have to do more with less resources—both money and people."

Cindy likes the flexibility of her current position, and the fact that she can more or less set her own hours. She also prefers "working for the government instead of industry." She says, "I feel our mission is more humanitarian . . . public health instead of money." She gets great satisfaction from the fact that she is able to have input into public health policy.

Her salary as a manager is slightly more than that of a chemist who remains in the laboratory, but as a government employee, it is less than that of an industry manager.

As for what she may do next, Cindy says, "To take the next step to go up, or even laterally and do something different within the FDA, I would need to relocate to Maryland. There is a lot of opportunity there, so it would just be a matter of me—and my family—deciding to go."

Advice

Cindy says, "Having a degree from a top school has really helped me. People do notice the school and remark on it. I think it can get you in the door when there are a lot of resumes. In fact, at one interview it was mentioned as a reason they brought me in for the interview.

"Definitely develop good communication skills—both written and oral. Also develop people and conflict resolution skills.

"You have to make your own opportunities. Nobody is going to look out for you, so be assertive about what you want for your own career."

Predictions

"I think the government is going to continue to be flatter; everyone wants fewer layers," Cindy says. "One problem with this is you sometimes get too many people under one first-line supervisor, and the supervisor gets swamped. There are about 30 people here, and, especially at review time, it is tough to keep up with all of them. We do have team leaders, but they are not allowed to do personnel-type work.

"Research and development in government is only going to succeed if we partner with industry and academics. We don't have much funding, so we find ourselves needing to work with others to assess new technology, but we also need to maintain objectivity and independence since we are a regulatory agency. It is a constant balancing act."

Veena Chorghade

Consultant and Vice President
Chorghade Enterprises

BS, Chemistry, 1978, Osmania University, Hyderabad, India
MS, Biochemistry, 1980, University of Bombay, India

Current Position

Veena is self-employed, providing consulting services to the pharmaceutical industry in the areas of regulatory compliance and overseas market opportunities. A major part of her work is writing technical reports, such as the comprehensive sections of chemistry and manufacturing controls (CMC) reports, development reports, stability reports, Standard Operating Procedures (SOPs), Corrective and Preventative Actions (CAPAs) and other regulatory documents for biopharmaceutical companies. In order to prepare these documents, she must perform statistical analysis, set up data tables, and analyze the results of her analyses. In some cases, she must critically evaluate the available data and write recommendations for future action.

She also advises companies on market technologies and the capabilities of overseas laboratories, particularly those in India.

In addition, Veena conducts current Good Laboratory Practices (cGLP) and current Good Manufacturing Practices (cGMP) training and assists with implementation of these protocols in both academic and industrial laboratories.

Since she is a consultant, Veena has a great deal of flexibility in her working day. She usually works 16 to 24 hours per week, spread out over 3 days. Veena says, "I try to finish my tasks for the week by Wednesday and send the reports to my clients by Thursday."

Career Path

After obtaining her master's degree, Veena spent nine years working at the Vijaya Pathology Laboratory, conducting radioimmunoassays and other analytical investigations. Over time, she advanced to management, and by the end of her time there, Veena was supervising 15 technicians.

In 1990, she moved to France and spent two years screening drug candidates for anti-viral activity at the Institut de Bacteriologie et Virologie.

She then moved to Chicago and obtained a position as an analytical chemist at Abbott Laboratories where she not only developed new protocols but performed all her work under current cGMP and current Good Laboratory Practice cGLP. These regulations, enforced by the U.S. Food and Drug Administration for the pharmaceutical, biologics, medical device, and diagnostics industries, ensure that medications meet quality standards. Veena also conducted audits to ensure compliance with the various regulations, which increased her familiarity with them.

After five years at Abbot, Veena relocated to Boston and responded to a Genetics Institute (now Wyeth Biopharma) advertisement. She was promptly interviewed and offered the position of bioanalytical chemist. She recalls, "It was a very pleasant and textbook-fashion job-application process." Her responsibilities included developing new procedures for various analytical methods for bulk drugs, drug products, or in-process samples for quantitation and further analysis, and providing analytical support for projects during development, pre-Investigational New Drug Application, and cGMP campaigns, and optimizing and validating various analytical procedures.

She also had to analyze the data from all the experiments and write methods, SOPs, validation reports, and so on. As her career progressed, she was put in charge of a wider variety of analytical support and was given more and more responsibility for statistical analyses, SOPs, methods, and various report-writing tasks.

However, her commute to work was about an hour each way—during snowstorms it could take two to three hours. Veena recalls, "I was very stressed because of the commute. At that time my son was about nine years old, and he did not want to go to after school programs anymore. I asked the company if I could go from 40 hours a week to 24 hours a week but was not able to do so. They did offer flexible hours, where I could go in at 6 a.m. and leave by 3 p.m.

I tried that for a few months, but that was not easy. I decided I had to give up my full-time job and do something part-time. I knew I could not be a part-time analytical chemist in a lab, so I started looking at other options."

Initially, she started out helping her husband, who was already a consultant, with his work on regulatory projects. She then started getting work by subcontracting from other consultants who needed help to meet their deadlines. From there, she expanded into taking on her own projects in the analytical area, and doing regulatory-submission projects for some of the smaller biotech companies in the area.

Veena's expertise, which she now sells to her clients, came from a decade of experience in the pharmaceutical industry. Advanced courses on pre-approval inspections for new drug–application and product license–application submissions, HPLC, biological safety for handling biohazardous materials, radiation safety, and clinical applications of radioimmunoassay techniques prepared her well for this career path. She says, "My experience in both big and small pharmaceutical companies has been crucial. Expertise in FDA guidelines and regulations, analytical thinking and statistical analytical expertise, writing skills, computer savvy, and so on, all come into play. In fact, virtually all aspects of my bioanalytical skills are required for this work."

Veena has now been on her own for almost three years. Most of her projects last 4–6 months. In most cases, she meets with the client about once a month and works on the projects from her home office the rest of the time. She spends about 75% of her time reading, analyzing, compiling, and writing reports and about 25% in meetings, on conference calls, and so on.

Veena loves what she is doing. "I love working from my home office, and plan to continue indefinitely," she says. "With every project, I get more experience, get more people willing to listen to me, and can increase my consulting fees—allowing me to make more than I ever received as a bench chemist. However, I do miss the typical office environment and meeting friends on a regular basis. Working from home can be very isolating."

Advice

Veena suggests, "Be flexible, and learn as much as you can, because you never know what you'll need to know to take advantage of new opportunities.

"The challenge of the independent consultant is to keep the projects ongoing. Selling yourself is a challenge and is a skill you must learn if you are to survive."

Predictions

"There is very little chance that the amount of regulation will decrease," Veena observes, "so companies will continue to need help with regulatory submissions. I predict a very bright future.

"I feel that a lot more men and women with experience in many of the regulatory aspects of pharmaceutical companies envision working from their home offices for various reasons. Technological innovations in communications, computers, and so on, will certainly aid in consultants like me being able to find opportunities to work for companies in any part of the world."

Laura M. Rosato

Global Product Regulatory Stewardship Leader, Honeywell Electronic Materials Honeywell International

BS, Biochemistry and Biophysics, University of Pittsburgh, 1981
MS, Biochemistry, University of Pittsburgh, 1985
PhD, Industrial Toxicology, University of Pittsburgh, 1990
MBA Certificate Program, University of Pittsburgh, 2002

Current Position

Laura Rosato is the global product regulatory stewardship leader for Honeywell Electronic Materials. As such, she is responsible for supporting the product-safety and product-stewardship issues related to toxicology, industrial hygiene, and other safety issues for new-product development and manufacturing. Product stewardship refers to the inclusion of safe and proper waste disposal measures in the distribution chain of an industrial product and to product sustainability.

Her job involves providing development, implementation, and leadership support for global regulatory compliance, and product-stewardship management for the electronic materials business. Her responsibility includes ensuring that all the company's products and manufacturing processes comply with regulations such as the federal Toxic Substances Control Act and similar regulations in other locations and at other levels of government. Laura also oversees registration and notification of products to appropriate authorities globally, ensuring that people use the material or product appropriately; hazardous communication (material safety data sheets and labels); brochures; interactions with the regulatory agencies or customer regulatory-affairs personnel; and advocacy efforts through trade associations.

Typical tasks include interacting with product-development personnel around the world, performing risk assessments on new or existing products (and for new applications), determining the global regulatory status of a specific product, developing safe levels of exposure and safe-handling practices for manufacturing and use of materials, and interacting with government or trade associations.

She says, "I am responsible for the products created by Honeywell Electronic Materials from cradle to grave. No day or week is the same. One day I will be working on developing and implementing programs on a specific product issue in Korea. I may interact with a trade association or perform a risk assessment on a product that is progressing through the product-development process. I do not have a 40–hour work week, but I get wonderful experience interacting with people from different cultures all around the globe. I extensively use online meetings for electronic collaboration, and sometimes travel to the source to work on issues. I live in the eastern time zone and need to interact with China, Korea, Singapore, and so on, so sometimes I have extremely long days and evenings."

Career Path

Laura received a BS in biochemistry and biophysics in 1981 and an MS in biochemistry in 1985, both from the University of Pittsburgh. She chose toxicology for her graduate work because she "wanted to be involved in a more applied field, one that required the use of a broad variety of skills." She notes, "Toxicologists have a wide range of areas they can be involved in—laboratory work, regulatory activities, expert witness for occupational diseases, setting occupational exposure limits, consumer products, chemicals, metals, pharmaceuticals, and so on." In 1990, she received a PhD in industrial toxicology, specializing in inhalation toxicology and industrial hygiene.

Upon graduation, she went to work at Procter & Gamble (P&G), where she planned and executed safety-testing programs to support new-product development and reformulation of existing products. She also performed exposure and risk assessments on various products to ensure they complied with all appropriate regulations, and worked with the Cincinnati Poison Control Center on issues concerning P&G's products.

After a few years, Laura realized that, as 1 of over 200 toxicologists at P&G, she faced tremendous competition for few professional advancement opportunities. She started looking for a smaller company where she would have greater responsibilities immediately. She wanted to find a company that was more chemically oriented, to move away from consumer products.

Laura applied for a regulatory-toxicologist position at Quantum Chemical Corporation (now known as Millennium Petrochemicals) and moved there in 1992. While there, she managed the company's toxicology and safety programs. Laura monitored regulatory developments that affected the company's products, and made recommendations to upper management on how to address changes. This meant tracking changes at the regional (state), national, and international levels. She went from being 1 of 200 toxicologists to being the only toxicologist at the company.

Laura recalls, "I enjoyed becoming intimately involved in solving problems and interacting with people from many different backgrounds. Many times, working in the laboratory you just supply data to another person or group and may never find out what happened with the information. In the regulatory-toxicology area, you are part of the solution and shaping the way a regulation is brought forward. In moving out of the lab into the regulatory area, I expressed an interest and read and took as many training classes as possible to move me in that direction." Laura also "had the opportunity to participate in many trade associations in Washington, DC, and to represent the company globally." She says, "This happened as soon as I took the job at Quantum Chemical. At Procter & Gamble, that may not have happened until 5, 10, or more years or more down the road."

After a brief period at the American Chemistry Council (formerly the Chemical Manufacturer's Association) in 1998, Laura moved to Karch and Associates (which was purchased by BBL Sciences in 2000), a consulting firm specializing in toxicology, epidemiology, and risk assessment. There, she analyzed data from occupational exposures to various chemicals from manufactured products or manufacturing facilities and determined if the disease or injury an individual or group claimed was plausible, based on scientific studies. The products studied ranged from materials containing silicon, to those used in the production of spark plugs or pharmaceuticals, to those containing heavy metal or allegedly causing neurotoxicity in children, and more. Her work was mainly in defense of the chemical and pharmaceutical industry.

While at Karch and Associates, Laura also acquired management skills by overseeing seven scientists. She learned valuable business skills by "doing"—she developed marketing strategies for bringing in new clients and created long-term strategic plans.

After Karch, Laura spent four years at the ALCOA Technical Center, learning more about regulatory interpretations and how to manage health and safety, including managing product safety and stewardship across multiple sites. In 2004, she moved into her current position at Honeywell International, where she is responsible for developing and implementing product-safety and product-stewardship management systems for 18 diverse facilities in eight countries. She actively tries to incorporate environmentally friendly processes wherever possible.

Laura's career path is probably not typical. People who start out in toxicology usually remain close to that discipline, but Laura decided to branch out into activities with a more regulatory emphasis. She wanted to interact with people and be more involved in the decision-making processes than she was able to when doing purely lab work. She recalls, "I enjoyed chemistry, biology, and toxicology and wanted to apply the concepts in a non-laboratory setting. The regulatory or product-stewardship arena allowed me both to use both my breadth and depth of experience and knowledge and to interact with a wide variety of people and apply skill sets developed in different disciplines."

Some of Laura's employers gave her opportunities to attend outside training in the various areas, but she also participated in community service and obtained practical leadership experience through the Girl Scouts and other nonprofit organizations. In fact, she was recognized by the Girl Scouts for her leadership as a Woman of Distinction in 1997.

Laura says, "The solid educational basis in chemistry, biochemistry, and toxicology have given me the foundation to learn other areas. In every position, I have learned as much as possible about that area. I have worked in the consumer-product, chemical, and metal industries in the areas of toxicology, regulatory affairs, industrial hygiene, product safety, product stewardship, and health, safety, and environment areas. I currently support a business at Honeywell—Electronic Materials—that has a very broad list of products ranging from mining ore to high-purity metals (like titanium, used in sapphire wafer manufacturing), dielectric and dopant materials (materials used on the surface of wafers that allow the layering of circuitry), high-purity chemicals, thermocouples, interconnected packaging solutions, all the components inside your electronic equipment that allow for heat dissipation (high-purity metal screws, wires) and so on. It is a very broad business. My experience in working with and understanding the chemistry of raw materials and the regulations that govern their use has been one of the ongoing themes in all the areas."

Laura loves the flexibility of her current position and the fact that she is always learning new things because regulations are always changing. She gets to interact with technologists, regulatory professionals, other product stewards, government and trade-association professionals, toxicologists, chemists, business professionals, R&D personnel, sales and marketing professionals, customer service people—all levels of the organization, including the company president. Her least favorite part of the job is working long hours and having to wear multiple hats.

As for her future, Laura is interested in the executive MBA program at the University of Pittsburgh. She wants to incorporate more business sense into the daily decision-making processes. She is not averse to moving again if an interesting opportunity presents itself.

Salaries for regulatory-affairs professionals, particularly those with toxicology skills, are significantly higher than for bench chemists.

Advice

Laura says, "If you have a strong chemistry understanding, you can tackle any of the scientific fields. Chemistry knowledge is power. If you can understand the chemistry, you can be a valuable resource to the product-development process. You need to know chemistry to understand the Toxic Substances Control Act, for example, and other global requirements.

"With the consolidation of chemical companies in the United States in the 1990s, it is more important than ever to have a broad background and not be narrowly focused in one area. Companies want individuals who can cross boundaries into many different areas. With a broad background, you have better perspective on issues because you can analyze the impact across various disciplines, and communicate with people at various levels."

Predictions

Laura believes that the regulatory area will continue to grow. She advises, "Learn to read and understand regulations. Those who are able to keep up with the new regulations and changes in the world marketplace will succeed.

"Virtually every industry is becoming increasingly global, and virtually all companies as well are becoming more global. Learning another language—especially Japanese, Chinese, or another Asian Pacific language—will let you get in on the biggest growth in the world."

ADDITIONAL RESOURCES

American Chemistry Council (www.americanchemistry.com/s_acc/index.asp) represents the leading companies engaged in the business of chemistry

American Industrial Hygiene Association (www.aiha.org)—dedicated to health and safety in the workplace, community, and environment.

Federation Internationale Pharmaceutiques (www.fip.org) is a global federation of national organizations of pharmacists and pharmaceutical scientists

Food and Drug Administration (www.fda.gov) is responsible for protecting the public health

Food and Drug Law Institute (www.fdli.org) provides high-quality education in the regulatory field and a neutral forum for the generation of ideas and discussion of law and public policy for its legal, policy, and regulatory communities

Management Sciences for Health (www.msh.org) seeks to increase the effectiveness, efficiency, and sustainability of health services by improving management systems, promoting access to services, and influencing public policy

Marasco, Corinne A. 2002. Careers for 2002 and Beyond: Regulatory Affairs. *Chemical and Engineering News.* (December 2): 75–80.

Office of Personnel Management (www.opm.gov) posts government opportunities and job benefits

Regulatory Affairs Professionals Society (www.raps.org) posts a listing of degree and certification programs

Society for Hazardous Chemical Communication (www.schc.org) is a professional society organized to promote improvements in hazard communications in the chemical industry

Society of Toxicology (www.toxicology.org) promotes the acquisition and utilization of knowledge in toxicology and aids in the protection of public health

World Health Organization (WHO) (www.who.int) is a United Nations agency specializing in health issues

7 Chemistry and Public Policy

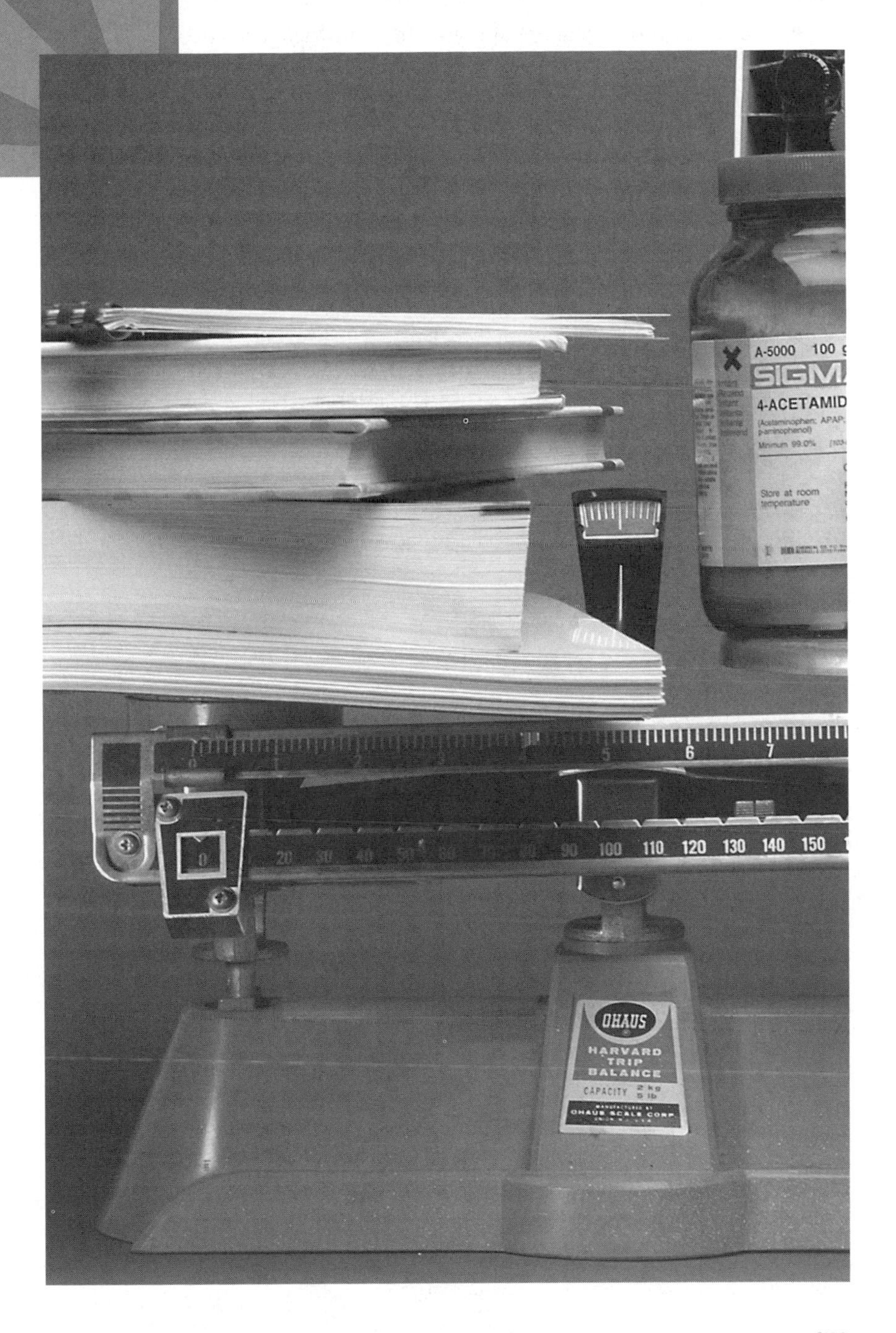

FOR THOSE WHO enjoy the application of science to the world around us, or who are passionate about improving the world we live in, a career in science policy may be ideal. No longer pursing the discovery of new knowledge, those who work in this field are applying that knowledge to the world.

Science policy often involves translating scientific findings into recommendations for action. In this profession, one provides governments with assessments and recommendations based on the best available scientific information, so that government officials can make sound policy decisions. The advisor must remain relatively neutral, or true to the ideals of the organization for which information is being collected, as appropriate. The job can involve coordinating the actions of groups or organizations that have very different interests.

Public policy involves working at the interface between science and policy. It requires thinking about the big picture, being able to identify future trends and to think strategically, and constantly being aware of one's obligation to the taxpayers or whoever is paying one's salary. The ability to think quickly, be flexible, and accept rapid changes in focus and direction are crucial to working in this field.

The ability to think analytically and critically can be more important than expertise in a particular scientific field, but having expertise in a field of science, which brings with it a fundamental understanding of a scientific topic, is also crucial.

* * *

THE traditional picture of a scientist is that of one who works in a laboratory, discovering new ideas and products or creating new materials. But in order for those discoveries to improve the lives of people, without having undesirable effects, policies must be implemented carefully. Making decisions on the use of new technology requires significant amounts of research into the technology and its possible applications, effects on other areas, and so on. Legislators do not have the time to do this themselves, so they hire advisors to keep up on specific areas and provide them with the facts they need to make informed decisions.

Science policy is a much broader field than most people realize. While budgeting and regulatory issues are often covered in the press, this field also includes the government's role in developing technology, coordinating various technologies toward a common national goal, developing and decommissioning weapons for national defense, developing drugs, ensuring food safety, establishing bioethics, and much more.

Many of those who make the decisions about how to implement technology policies are not experts in those technologies. They must rely on the advice of others, whose job it is to keep abreast of the latest developments and the possible social and economic effects of those developments. The amount of contact with people and access to information can be overwhelming, and developments in information technology have served only to increase the flood of information. In many cases, the job is one of filtering—selecting only the important bits of information to pass along to each interested party.

There are all sorts of people and organizations who regularly need advice on science and technology issues. For example, some embassies employ science and technology counselors (or advisors) to gather information and keep themselves and their counterparts at home updated on the state of technology in their host country, to set up collaborations between their home country and the host country, and to develop and nurture relationships between scientists and decision makers in both countries. When dignitaries from the home country come to visit, the advisor briefs them on the state of science in the host country. This involves following a wide variety of scientific areas, through publications, conferences, and networking; being constantly on the lookout to learn more; and, not least of all, being willing to live in a foreign country. The larger an embassy is, the more science and technology advisors it is likely to have, and the more specialized each one can be. These positions do not necessarily require a senior person with a lot of experience, just a reasonably broad science background and expertise and some international experience.

For those interested in gaining some experience in this area, the American Association for the Advancement of Science (AAAS), American Chemical Society (ACS), and American Physical Society (APS) all coordinate science policy fellowships that allow people to spend time (typically one to two years) working on Capitol Hill learning about how the government understands science. There are also congressional fellowships that allow scientists to spend a year working in a congressperson's office or working for a particular committee. Other programs include the Presidential Management Internship Program and White House internships, in addition to management courses in forestry or public health. Participation in any of these is a good way to see what working in science policy is like, and people can apply for these programs at any stage in their professional career. One should investigate the options carefully, as most programs are very competitive, and some fellowships are unpaid. However, the networking connections one makes in these positions are invaluable.

Many professional societies are interested in hiring people who have had experience working on Capitol Hill and who thus have connections and can help advance their particular agenda. Positions such as director of public affairs, federal grants manager, and others benefit from someone with a personal understanding of how things work inside the federal government.

The policy area also involves a lot of negotiations, complications, and adaptations as the many factors and influences that contribute to policy formulation shift. One may be called on to work in an unfamiliar cultural setting, either in a foreign country or in a setting in which one has different ideals and goals from other scientists and peer groups.

In some positions, depth of knowledge is not as important as breadth of knowledge about the capabilities of a particular country, perhaps for future collaborations; in other cases, in-depth reporting a particular technology and its implications for the future is required. Knowing which is required, and when, is one of the essential skills for a career in public policy.

Profiles

Dennis Chamot

Associate Executive Director, Division on Engineering and Physical Sciences
National Research Council, National Academies

BS and MS, Chemistry, Polytechnic Institute of Brooklyn (now Polytechnic University), 1964
PhD, Chemistry, University of Illinois, 1969
MBA, Wharton School, University of Pennsylvania, 1974

Current Position

Dennis Chamot works for the National Academies, which is comprised of four organizations—the National Academy of Sciences, the National Academy of Engineering, the Institute of Medicine, and the National Research Council (NRC). These groups bring together experts in all areas of scientific and technological endeavors, who serve pro bono to address critical national issues and advise both the federal government and the public.

Dennis is the associate executive director of the division on engineering and physical sciences for the National Research Council. The basic "business" of the NRC is to provide objective, credible, scientific and technical advice, primarily to federal agencies. With a very few exceptions, all NRC reports are public. As a member of management, Dennis is responsible for ensuring that the process produces good reports, and oversees the business aspects of the operation. Dennis says, "On occasion, I have had to resolve some personnel issues, establish better financial oversight within a divisional unit, or provide routine oversight for several studies that were under way. I have been called on three or four occasions during my tenure at the Academies to serve as acting board director for boards in difficulty, while still retaining my other responsibilities."

Dennis has also helped develop new business opportunities for the NRC, particularly for his division. He was instrumental in helping organize an ongoing roundtable discussion on biomaterials and medical devices, and another roundtable on critical infrastructure. He has been involved in many preliminary discussions with potential sponsors that have led to a variety of studies by several boards in his division.

Dennis undertakes a wide variety of activities, which include overseeing periodic external reviews of the 14 boards in the division, interacting with board directors and staff about the development of new projects and the conduct of existing studies, and overseeing bias and conflict of interest procedures for all the boards and study committees in the division. He has meetings with current or potential study sponsor representatives and frequent meetings with his boss, the executive director of the division on engineering and physical sciences.

Dennis notes, "Every day is different. If I don't have an early meeting to go to directly, I check e-mail messages to see if there are any urgent issues to deal with. The rest of the day would be a mix of activities and meetings.

"While the National Academies are private, nonprofit organizations, our basic charter—issued by Congress during Abraham Lincoln's presidential administration—calls upon us to provide scientific and technical advice to the U.S. federal government. The boards in my division provide advice broadly across the U.S. government, from the National Science Foundation to the National Institute of Standards and Technology (NIST) to the Department of Energy and Department of Defense, and simultaneously to the American public. Many of the issues we are asked to deal with are of great importance to the welfare and security of the country, and I derive great satisfaction from contributing to this effort. I also very much enjoy meeting and working with some of the top scientific and engineering leaders in the country."

Career Path

Dennis attended Stuyvesant High School in New York City and had already picked chemistry as his major interest when he applied to college. He pursued his chemistry straight through to the PhD. He recalls, "As with many other students who follow that path, and having thoroughly enjoyed my teaching duties, I thought I would like to follow a university career. However, I was not eager to devote another couple of years to a postdoctoral position."

Dennis instead went to work at the DuPont Experimental Station near Wilmington, Delaware, in 1969 as a research chemist. He worked in the organic chemicals department performing organic synthesis. "One of my earliest pleasures there was noting that the taxes taken out of my first monthly paycheck were more than my gross monthly income as a graduate student," he says.

After a couple of years, Dennis's entire work group was transferred to a plant site across the river. While the commute was the same, the physical and intellectual atmosphere at an old plant site was quite different from the campus-like environment of the Experimental Station. Dennis almost immediately thought about making a change, and recalls, "I realized that moving to another company would not necessarily satisfy me—DuPont at the time was certainly no worse a place to work than the other major companies. I just did not feel that this was the life for me." Dennis was also very much affected by the trauma and emotional problems suffered by friends and coworkers who lost their jobs because of a recession and through no fault of their own.

He decided he needed a significant change, so in 1973 he left the company to attend Wharton School full-time. He obtained his MBA in about 18 months.

While at DuPont, Dennis had become active in the local section of the American Chemical Society (ACS). He edited the local section's publication, the *DELCHEM Bulletin*, and was involved with the Younger Chemists Committee of the national ACS organization. While at a national ACS meeting, Dennis had heard a talk by the head of a council that later became the professional employees department of the AFL–CIO (American Federation of Labor–Congress of Industrial Organizations). When it came time to look for a new job, that speaker was one of the people Dennis contacted. This resulted in his being hired to work in that department. He recalls, "I had been active in the professional-relations movement in ACS for several years, and the fact that my MBA area of specialty was in personnel and labor relations was to key my being hired."

Dennis worked in the department for professional employees of the AFL–CIO from 1974 through 1993. The AFL–CIO is a federation of most major U.S. labor unions. Over the years, Dennis had various titles as the number-two person in the department for professional employees. Dennis's final title at the AFL–CIO was executive assistant to the president of that department. Throughout his years there, regardless of his title, his job involved primarily researching, writing, public speaking, and representing the organization at various meetings or on other organizations' committees, particularly at the U.S. Department of Labor, the National Science Foundation, the National Institute for Occupational Safety and Health (NIOSH), the Council on Competitiveness, and others. His department served a large group of unions that dealt with professional and white-collar employees—actors and musicians, nurses, teachers and professors, librarians, office workers, and scientists and engineers, among others. Their department supplied information, data, and analyses, and provided workshops, organized internal committees, and offered lobbying services to the affiliated unions.

One of Dennis's major interests was looking at the effects of new computer and telecommunications developments on job design, employment levels, and competitiveness issues, especially as they affected white-collar employees. Over several years, he wrote articles and delivered talks on these subjects, both to the union and other audiences and organized a couple of national conferences and several workshops on these subjects. Dennis also oversaw the preparation and publication of a review of union contracts dealing with technological change and related issues.

He became recognized as a labor expert in that field, and as such attended meetings and workshops with various groups as a representative of the AFL–CIO. Dennis says, "I have continued to be interested in the subject, especially with the recent expansion of globalization of R&D and other professional-level work, although I now look at it from a broader national perspective, not just from the official union position."

Dennis feels his chemistry training was very useful in several ways. He notes, "First, I brought with me knowledge of the world of scientists and engineers and an understanding of what their workplace and career concerns were. Second, the analytical skills I developed in studying chemistry served me very well when I became much more involved with gathering and analyzing economic and employment data—one still needed to be concerned about accuracy, the need for evidence to support arguments, analytical rigor, and the ability to clearly communicate conclusions. It was essential to be able to write clearly and to develop a good public-speaking style. Third, because of my understanding of science and technology, I developed a particular interest in studying the effects of new technologies on work design, employment, and competitiveness issues. I wrote articles, presented talks at professional society meetings, and participated in committees at the National Academies, the National Science Foundation, and the U.S. Departments of Labor and Education, and also developed workshops and written materials for various labor groups. As a result of these activities, I achieved some broader recognition and was elected a fellow of the American Association for the Advancement of Science—rather unusual for an employee of the AFL–CIO."

Working as a professional staff member within the labor movement was much less lucrative than working in industry, but the pay was comparable to the remuneration an assistant or associate professor received. On the other hand, there is a great deal of personal reward in helping shape public policies and in offering assistance to organizations whose function is to help people better their lives.

Dennis says, "In general, my job at the AFL–CIO was atypical for a chemist although personally satisfying. Similar opportunities requiring similar skills and making use of one's science background would include working for public-interest groups involved with the environment or consumer affairs, or for political groups with which one sympathizes."

Dennis remained at the AFL–CIO for almost 20 years and for the most part found the work very satisfying, both professionally and ethically. When it came time for a change, he met with the president of the National Academy of Engineering (NAE), who had gotten to know Dennis through his membership on several National Research Council committees and speeches he gave at a couple of NAE conferences. This meeting started a chain of events that led to Dennis's being hired as the associate executive director of the Commission on Engineering and Technical Systems (CETS) at the National Research Council.

CETS merged a few years later with another internal commission to form the Division on Engineering and Physical Sciences (DEPS), a unit about twice the size of CETS, both in terms of its budget and the number of employees, and Dennis became the associate executive director of the larger division. The DEPS comprises 14 boards that deal with public policy issues in many areas of science and engineering. Dennis is part of the senior management team that oversees the work of the boards and their study committees, involving over 100 employees, an annual budget of approximately $25 million, and several hundred volunteer committee and board members. Much of the division's work involves assembling expert committees to study questions of public policy that have a major technical component. The task may involve reviewing specific government programs, making recommendations for prioritizing areas of research requiring federal funding, informing various agencies of state-of-the-art developments in basic and applied science, or advising on policy options. Dennis is involved in reviewing the boards' programs and finances, developing strategic plans, adjudicating questions of bias and conflicts of interest involving committee members, developing program activities, and so on. The work is interesting and varied, and he gets to meet a lot of important people. While the National Academies are not part of the federal government (they are private nonprofit organizations, tracing their history back to the establishment of the National Academy of Sciences in 1863), much of their work is federally funded. Even so, the government does not control the work or the content of reports. The independence of the National Academies is essential to their credibility and influence.

Working at the National Academies is not as much of a stretch from working as a scientist as working at the AFL–CIO was, because the major issues studied have a strong scientific or technical focus. But similar skills are involved: one needs to be able to communicate clearly, both verbally and in writing, and be able to present arguments logically and show supporting evidence. On the management side, good interpersonal skills, an ability to plan and direct work, and familiarity with budgets are also important.

Dennis says, "I see myself as having had careers in three different areas up to this point—industrial chemical research, the labor movement, and science and technology policy. People chuckle when I refer to this as a typical career path, as it clearly is not. But then I point out that with some talent and ability, and an excellent chemistry education, many things are possible."

Advice

Dennis advises, "Over time, develop a good broad science and technology background and get involved in public policy issues with a heavy technical component."

Predictions

Challenges continue to draw professionals into this field. Dennis notes, "The biggest problems facing the human race today are poverty, disease, and ignorance. There are exciting challenges in developing new medicines and medical devices; in reducing the costs of housing, food, and transportation; and in developing useful educational technologies. All these areas are dependent upon or enabled by advances in chemistry and materials. The basic science will continue to develop in response to the curiosity of good scientists, no matter what, but these other challenges require more.

"Personally, I am excited by the enormous advances being made in biology and medicine as a result of our developing greater chemical understanding of the underlying structures and reactions. There is still much to be done in understanding and applying the chemical basis of biology. Another exciting area is nonbiological specialty materials, which enable advances in a broad range of applications."

Lee M. Nagao

Senior Science Advisor
Gardner, Carton, & Douglas, LLP

BA, Chemistry and Biochemistry, University of California, San Diego, 1990
PhD, Chemistry, Yale University, 1998

Current Position

Lee Nagao is a senior science advisor at a national law firm, which means she represents clients to U.S. and international regulatory agencies and standard-setting bodies for the regulation and control of pharmaceutical-drug products. These groups include the International Conference on Harmonisation (ICH), the European Medicines Agency (EMEA), and the International Organization for Standardization (ISO). This position lets her combine her interests in science, writing, and policy.

Lee works with clients, most of whom are scientists at pharmaceutical companies, to develop data and science-based reports and guidelines for publication and/or submission to various regulatory agencies and advises clients on interactions with these regulatory bodies. The proposals she prepares cover a wide range of pharmaceutical science and manufacturing issues including toxicology and safety, analyses of leachates (chemical compounds that seep into the drug product from the product container materials), current Good Manufacturing Practices issues, and foreign-particle analysis. She also manages scientific and regulatory working groups consisting of scientists from industry, the U.S. Food and Drug Administration, and academia. Finally, she provides scientific advice and explanations to the attorneys with whom she works.

Lee says, "A large portion of my work involves helping clients develop data-based position papers and presentations for regulatory agencies such as the FDA. For instance, I am involved with a group of scientists from industry—some of whom are clients, the FDA, and academia who are developing science-based recommendations to the industry on how to control leachates in inhalation and nasal-drug products. Leachates—or 'leachables'—are chemical compounds that may be pulled into the drug product from the surrounding packaging or product-container materials or by the action of the drug-product solvent or formulation. I am managing the work of this group, and also contributing to the analyses of data, development of experimental protocols to acquire the data, drafting of the recommendations, and development of strategic plans that facilitate the functioning and effectiveness of the group. We hope that the recommendations we develop will contribute to the development of regulatory guidelines in this area."

Lee is on the road 10–20% of the time, meeting with clients and auditing chemistry labs both domestically and internationally. She says, "A quality-assurance audit of a pharmaceutical-testing laboratory is performed to ensure that the lab has incorporated and acts upon procedures and systems that allow it to operate at a specific level of quality, as expected by the pharmaceutical company and the FDA. An audit may include checking the labs and documentation to make sure that standard operating procedures exist and are followed, and that all documentation and lab procedures adhere to good manufacturing practices and/or good laboratory practices, commonly called GMP and GLP."

Lee's typical day is extremely busy, usually filled with teleconferencing with individual clients and working groups, drafting papers and reports, and attending internal meetings or in-person meetings with clients.

Career Path

In graduate school, Lee studied solid-state physical organic chemistry. After obtaining her PhD, she was interested "in writing, in exploring further the ramifications of science on the public and on policy matters, and in engaging scientific matters in the public realm rather than just in the laboratory environment." She chose to pursue science journalism and spent a year doing freelance science writing. During this time, she was a contract writer and researcher for National Geographic Television.

Lee really wanted to combine her interests in science, public policy, and writing and to do more research and policy-based work. A friend from graduate school heard about a law firm that was looking for a chemist to serve as an advisor and recommended Lee. She jumped at the opportunity, went for the interview, and was hired.

Lee recalls, "I was somewhat 'off the beaten path' already in taking the science-writing route after graduate school instead of going into academia or industry. I was able to take advantage of this opportunity since I was looking for a job change, but I had also already launched myself into the alternative-career realm. I feel that this paid off in allowing me to expand my experiences and skills and be in the right position to find my ideal career."

Lee's responsibilities have increased in the five years she's been at the law firm. She is now engaged in a much wider variety of activities. In addition to writing policy and technical papers and advising on scientific matters, she spends a significant amount of time in a management role, helping lead and coordinate activities in which the firm's clients are involved. She explains, "This may involve coordinating the activities of testing labs or managing scientific or industry working groups. I also represent our clients to regulatory or other policy-making agencies and get many opportunities to present at conferences."

Lee loves her work, and notes, "I've found a career that lets me explore my interests—science, policy, and writing—but also constantly introduces me to new situations and new skills such as management, public speaking, and business. My colleagues are very smart, stimulating and are a joy to work with. I've recently had a child and my work environment is extremely supportive of my situation, giving me the flexibility to work at home several days a week and letting me craft my own schedule."

Lee interacts daily with scientists (chemists, statisticians, engineers, medical doctors), regulators, attorneys, paralegals, administrative assistants, and many others. The job allows for personal growth, in that Lee has the "opportunity to manage others, and to interact with government officials and people outside the scientific community." She adds, "It also allows for knowledge growth in areas outside what I studied in graduate school, such as business, biology, pharmacy, policy, statistics, law, patents, and so on."

The skills that Lee finds most valuable include a broad scientific understanding; good writing skills; good speaking skills; the abilities to prioritize, listen very carefully, take lots of information and condense it down to just the key points, facilitate discussions among people with disparate views, understand opposing points of view; and most of all patience.

Many aspects of Lee's chemistry training are useful day to day, "including general training, such as the ability to create and plan projects, manage projects, and see them through to a resolution; the ability to write and present issues succinctly; general knowledge of chemical and physical laws and theories; and a general comfort level and confidence when addressing or learning more about a broad range of chemical problems and issues."

The work also provides opportunities for creative application of scientific knowledge and writing skills, which Lee enjoys very much. Her biggest challenge is "developing the strength, patience, and savvy to work effectively with government agencies." Her only regret is that she didn't take more biology-based courses in graduate school.

Lee plans to continue growing in her current role and advancing with the firm.

Advice

Lee advises, "If you are really interested in this career, you should be sure to keep your background broad and not be afraid to jump into new things in order to expand your knowledge areas. Learn all you can about government and policy, especially the U.S. Department of Health and Human Services and the FDA, and the world of pharmaceutical development. Hone your writing skills!

"I recommend speaking with scientists in these careers or those who have made the change to an 'alternative' career. Usually, they are the most helpful and may have the right contacts to help you out."

Predictions

Lee says, "There are many good opportunities for scientists in this field, as many law firms are looking to expand their staff to include 'non-attorney' professionals who can give their firms flexibility and depth. This will most likely continue to be the case."

James G. Martin

Corporate Vice President
Carolinas HealthCare System

BS, Chemistry, Davidson College, 1957
PhD, Chemistry, Princeton University, 1960

Current Position

Jim is currently the vice president in charge of government relations for Carolinas HealthCare System, one of the largest publicly owned health care systems in the country. He spends most of his time lobbying, or making sure that the opinions and needs of his organization are represented in the government. He meets with government officials regularly and carefully tracks the progress of legislation and policies that affect his company's interests and those of all public hospitals and their patients.

Career Path

Jim's first job after graduate school was as an assistant professor of chemistry at Davidson College, where he taught and conducted research for 12 years.

While at Davidson, he successfully ran for a position on the Mecklenburg County Board of Commissioners. He was a commissioner for six years, serving as chairman for the latter part of his tenure there. He initiated the first environmental enforcement program ever in his state and led the adoption of the first countywide zoning in North Carolina. He also served as the president of the North Carolina Association of County Commissioners from 1970 to 1971.

Looking way a way to make more of an impact, Jim ran for a seat in the U.S. House of Representatives in 1972. He was elected and ended up serving six consecutive terms. During his tenure there, Jim served as a House Republican Research Committee chairman and as a member of the Ways and Means Committee.

The latter committee has jurisdiction over all taxation and tariffs, and other revenue-raising measures, in addition to a number of entitlement programs, including Social Security, unemployment benefits, Medicare, child-support enforcement, Temporary Assistance for Needy Families (a federal welfare program), and foster-care and adoption programs. This is an especially powerful committee, since, constitutionally, all bills regarding taxation must originate in the House of Representatives, and therefore all must go through this committee.

In 1983, James became the first elected official to receive the Charles Lathrop Parsons Award, given by the American Chemical Society for outstanding public service by an American chemist.

In 1984, he was elected governor of the state of North Carolina, becoming only the second North Carolina Republican governor to be elected in the twentieth century. He also became the second North Carolina governor to serve two successive terms. During his term in office, he especially supported education and commerce in North Carolina.

Jim recalls, "Each time I ran for a new public office, early polls showed me trailing badly, so the successful election returns were unexpected to many."

In 1993, he stepped down as governor and prepared to return to private life. He was offered a position heading the research lab at Carolinas Medical Center. His chemistry background allowed him to successfully direct the research of a number of biologists. This includes consulting, providing guidance, giving direction to needed resources, providing information on applying for funds, training biologists in regulatory oversight and Good Clinical Practices, administering finances, managing projects, and explaining institutional review board (IRB) procedures.

An IRB or independent ethics committee (IEC) is a group of experts who monitor and review research involving human subjects—generally biomedical or behavioral research—mainly to provide a check for ethical problems. They have the power to approve, require modifications to, or disapprove research projects. In the United States, IRBs are mandated by the Research Act of 1974, which requires them for all research that receives funding, directly or indirectly, from the Department of Health and Human Services (HHS). Today, many IRB reviews are done by outside, for-profit organizations.

In 1999, instead of retiring, Jim again left the research world to move into public policy, this time as a lobbyist. For this work, the most important factors are being friendly and having high-level contacts within the government.

Advice

Jim advises, "With regard to a political career, unless you are independently wealthy or famous, you should begin by working for others, either in political campaigns or volunteer organizations. When you are ready to run yourself, build a strong 'grassroots' organization."

Predictions

One trend particularly worries Jim. He observes, "Politics is becoming more polarized, and less civil, which is troubling."

ADDITIONAL RESOURCES

American Association for the Advancement of Science (AAAS) (www.fellowships.aaas.org) offers yearlong fellowships in science and technology policy

Association of Schools of Public Health (www.asph.org)

European Medicines Agency (www.emea.eu.int) evaluates medicines

Food and Drug Law Institute (www.fdli.org) provides education and a neutral forum for the generation of ideas and discussion of law and public policy

International Conference on Harmonisation of Technical Requirements for Registration of Pharmaceuticals for Human Use (ICH) (www.ich.org) brings together the regulatory authorities of Europe, Japan, and the United States and experts from the pharmaceutical industry in the three regions to discuss scientific and technical aspects of product registration

International Organization for Standardization (ISO) (www.iso.org) is a network of the national standards institutes of some 150 countries, with its central office in Geneva, Switzerland

National Association of Science Writers (www.nasw.org)

Office of Science and Technology Policy (www.ostp.gov) exists to advise the U.S. president on the effects of science and technology on domestic and international affairs

Union of Concerned Scientists (www.ucsusa.org) is an advocacy group using rigorous scientific analysis and innovative thinking to promote a cleaner, healthier environment and a safer world

8 Chemistry and Safety

SAFETY, IN THE most general sense, includes the safety and long-term health of employees, users, and neighbors, and the safety of the environment and the planet in general. This very complex and highly regulated area employs a huge number of people and involves maintaining a delicate balance between risk and frequency, and between time and cost.

Companies have always tried to minimize time lost because of accidents, and for a long time have also been reducing energy use and waste emissions. However, many companies now are committed to making processes that are sustainable—using resources in a way that meets current needs without compromising resources that future generations might need.

Working in this field provides a different set of challenges from laboratory research. Safety, of both people and the environment, involves carefully planning, executing, and monitoring complex activities to support the business process, not interfere with it. Both reacting immediately to short-term crises and making long-range plans can be required, in addition to collecting and monitoring samples on-site, performing in-office planning, and writing reports.

In the long run, chemists working in safety and related fields aim to inspire action and influence change for the better at the laboratory, corporate, and governmental levels.

* * *

As any good chemist knows, nothing is completely safe, and nothing is inherently risky. Water can be toxic (drowning), and explosives can save lives (nitroglycerin can be used medicinally as a vasodilator). Any process can be carried out in a variety of ways at many points along the safety continuum.

This large and complex matrix of short- and long-term health issues, regulatory-compliance issues, risk management, cost–benefit analysis, and so on, results in a large number of career possibilities for those with a chemistry background. Companies must comply with all laws, regulations, and requirements that apply to their industry and activities, and they must be aware of changes to those policies and adapt to the new requirements. This includes regulations at every level—national, state, and local—and can involve resolving disputes when multiple overlapping requirements conflict. In many cases, companies are audited to ensure they are in compliance, and many companies use consultants before an official audit to ensure they are properly prepared. Furthermore, safety officials must ensure compliance with all relevant regulations by complex, diverse groups with sometimes very different interests and motivations. Negotiation and consensus building are critical skills for these situations.

Keeping current on all the relevant requirements is a huge task itself, so training is a big part of safety-related careers. Managers and auditors must know what the current regulations are. Employees working "on the floor" must have the proper training for the tasks they perform and the environment in which they work. Often, regular refresher courses or continuing-education courses are required for work in a particular area. When planning training courses for staff, one must plan carefully, as there is a high activation barrier to overcome in getting people to attend safety classes.

Someone must record what the company has done, and someone must audit the entire process to make sure the records are accurate and complete. Record keeping and documentation are often the most important parts of the safety process. Analyzing the results of audits and recommending actions for improvement are also important ongoing processes. Nothing is ever perfect, but current practices should reflect the best methods possible within the existing constraints.

Occasionally, company officials must deal with advocacy groups or with the media, primarily when a crisis occurs. In these cases, the safety officials must be able to quickly provide accurate, complete data and help present it in the best possible light.

Keeping up with changing regulations is crucial; it is important to know what new requirements are coming so that the company can be prepared to promptly comply with them. Often, this means continuing education and certification are required. In some cases, it means learning completely new sets of regulations, such as the process safety-management requirements that were implemented in the United States in 1984. These requirements deal with managing the hazards associated with processes that use highly hazardous chemicals (toxic, reactive, flammable, or explosive), with the twin goals of maintaining a safe workplace and preventing chemicals' accidental release into the environment.

As regulations become more complex and compliance possibly more expensive, interesting ways to meet the need for continuing education are being developed. For example, Texas has outfitted a 41–foot trailer with a laboratory to provide water- and wastewater-treatment classes around the state. Instructors provide accredited classes in efficient, accurate, and safe laboratory-testing practices to the wastewater-plant operators and others who need them. Instead of having students travel to a central training location, the teacher takes the lab to them. Operators get hands-on training and keep their certification, while taking minimal time away from work. Texas is a large state with well over 22,000 wastewater operators to train and certify, so it makes more sense to take training to them.

In order to be competitive, companies must keep costs down. Health and safety are often seen simply as cost centers, and sometimes these functions are outsourced to the lowest bidder. Especially in these cases, safety managers must be able work with the company to convince them of the importance of safe practices and must be willing to defend their professional opinion of what the minimum safety requirements are, sometimes in the face of strong economic pressure.

Safety

Safety involves the physical well-being of the people working within a manufacturing plant or laboratory, the people using the end product, and those disposing of a product.

Obviously, a first step is to identify potential hazards and develop methods to control them. Virtually all chemicals can be safely handled after their hazards are properly evaluated. Safe handling may require appropriate personal protective equipment; training in the use of ventilation, engineering controls, and other equipment; or a knowledge of other procedures to reduce the risks to acceptable levels. Chemicals must also be identified, labeled, stored, and disposed of properly. Safety hazards must be identified and reduced at all places in the production cycle, during storage and use, and also during transportation.

A large part of safety work is performing risk management—measuring or assessing the various risks of a particular course of action—then taking steps to reduce or transfer the risk to others and implement systems to manage any possible adverse outcomes. For example, scheduling preventative maintenance on equipment reduces unexpected system downtime, but the risk of unanticipated downtime must be balanced against the cost of planned downtime.

Besides the materials with which people are working, there are more general workplace issues to consider, such as the presence of asbestos and biological and blood-borne pathogens, radiation and laser safety, drinking-water quality, the availability of first aid, hazard communications (HAZCOM or worker's right-to-know), air quality, hearing conservation, and anything else that might affect the personal safety of employees or visitors. All these must be monitored so hazards can be addressed.

Although removing or reducing risk is always the best idea, knowing what to do in case of an incident is imperative. Emergency plans must be in place describing what to do in case of accident in the laboratory or workplace, or in case of an incident outside the company. Safety officers make sure that personal protective equipment, including lab coats, safety glasses, fire extinguishers, eye washes, and showers are available and working at all times. They may be in charge of stocking other types of supplies, such as proper containers for chemical waste, or recycling facilities for particular items. After an incident, they may be responsible for investigating the cause of the accident and recommending preventative measures.

Safety can also mean dealing with security issues, such as ensuring that all areas are closed to access by unauthorized personnel and that human factors don't override security procedures.

Health

Health is a logical extension of safety and looks at the longer term effects of the work environment on people's physical and mental well-being. It is often difficult to say where safety concerns end and health concerns begin.

Health issues can involve safely using electrical systems, compressed gas, or flammables, and implementing lockout and tag-out systems to reduce the risk of incidents. Developing procedures to safely enter confined spaces (such as storage tanks that need cleaning), putting guards on machinery with moving parts, analyzing the potential for injury from hoist and overhead systems, and regularly monitoring air quality all fall within health concerns. Injuries, such as carpal-tunnel syndrome or other repetitive motion (stress) disorders are also health issues, requiring that someone attend to the ergonomics of the workplace.

More and more companies are integrating health and safety issues into their organizational values. This sometimes requires making behavioral and cultural changes in the organization.

Environmental Issues

Environmental issues concern the safety of the planet as a whole—making as small an impact as possible during all phases of business. Manufacturing facilities can have huge effects on the air, land, and water around them, and those effects must be minimized and monitored to ensure they are as innocuous as possible. Those interested in environmental management look at the fate of chemicals in the environment and the effects they have on the environment as they pass through. Both the chemicals and the environment they are in are changed by their interaction with each other.

Many companies are now using "green" chemistry, which tries to reduce and prevent pollution at its source. This means designing chemical products and processes that reduce or eliminate the use and generation of hazardous substances. For example, using newer, more energy-efficient lighting and motors in place of older ones can save energy. This change also helps the company's bottom line; waste disposal continues to get more expensive, so the less there is to dispose of, the better.

Air quality, storm-water contamination, and watershed (the region that drains into a particular river, lake, or other body of water) effects come into play when designing or operating a facility. Environmental managers address damage by designing abatement systems that minimize the amount of unwanted materials that escape into the environment. Filters and other devices are used to minimize the amount and impact of waste effluents, and these devices' performance must be monitored and, in many cases, reported to an authority regularly.

The hazards of transportation can sometimes be eliminated if a product can be manufactured at the site where it is needed.

It is not always possible to prevent environmental contamination. When contamination does occur, the pollution or contaminants must be removed from land, air, and waterways. This process is called remediation. The first step in remediation is to verify the existence (or absence) of contamination. Next one must identify what the contaminants are, how much is present, and whether they can or should be cleaned up. Sometimes determining exactly what the contaminant is and how long it has been there is crucial if one is to collect funds for cleanup from the responsible party. Finally, the best cleanup method must be chosen and implemented. The best method for decontamination depends on the physical, chemical, and biological properties of both the site and the contaminants. One technique, called bioremediation, uses the existing soil bacteria to decontaminate a site. This technique usually requires that the contaminant is a carbon source, that it is (or can be made) water soluble, and that there is a proper balance of nitrogen, phosphorus, and electron acceptors (for example, oxygen). Existing bacteria in the soil then metabolize the contaminant and clean up the site.

Work in the environmental-safety field requires knowledge of chemistry and of regulatory and compliance requirements. It may involve collecting samples in the field and/or testing them in a laboratory. It often involves explaining results to nonscientists (sometimes in court), so the ability to communicate complex ideas without using technical jargon is essential.

Industrial Hygiene

Every workplace has its own potential long- and short-term hazards. Industrial hygiene is the anticipation, recognition, evaluation, and control of workplace environmental factors that may affect workers' health, comfort, or productivity. Industrial hygienists take a more global look at the workplace, identifying and remediating potential hazards. In general, a laboratory-safety officer focuses on the practical matters of safety for the employee or student, such as training them in the safe handling of chemicals or inspecting laboratories and chemical-process work areas. The focus of an industrial hygienist is similar but more complex in that the individual is trained to look at workplace or laboratory conditions with a view to preserving the health and safety of the chemical-process workers, laboratory employees and technicians, and office workers as they do their work.

Industrial hygienists have the background, tools, and resources not only to recognize hazards but also to evaluate them and recommend control measures to preserve the health and safety of those exposed to the hazards. They do this by conducting regular inspections, taking air samples, analyzing hazardous constituents, and working with chemical, mechanical, and facility engineers to address relevant issues. A well-grounded knowledge of chemistry is of tremendous value, but the industrial hygienist also has training in engineering, HVAC systems, fume hoods, and air-quality testing (and result interpretation). They can make recommendations for changes and often have the authority to insist on changes.

Few schools have undergraduate degrees in this field, so most people get an undergraduate degree in chemistry, biology, or engineering then a master's or doctoral degree in industrial hygiene. After five years of experience as an industrial hygienist, one can take a comprehensive one-day examination to become a certified industrial hygienist; the process is similar in form and intensity to the one accountants undergo to become certified public accountants. Recertification is done on a five-year cycle and requires active professional involvement in the field and continuing-education courses.

Industrial hygienists are employed in public utilities, the government, labor unions, laboratories, hospitals, insurance companies, chemical and manufacturing companies, and many other places. According to the American Industrial Hygiene Association, the fastest-growing segment of the profession is the self-employed or consultant.

Profiles

Linda Wraxall

Criminalist Lab Safety Officer
California Department of Justice DNA Laboratory

BS, Zoology, with minors in Botany and Chemistry, University of London, 1964

Current Position

As a criminalist lab safety officer, Linda is responsible for the safety of more than 100 scientists in the DNA lab. She oversees training, chemical inventories, troubleshooting, and preventative measures to keep her coworkers safe, healthy, and productive. Linda takes her responsibilities seriously and will "cause trouble to get things done" when she thinks people or property are at risk.

Linda oversees safety training for the entire laboratory staff. She explains, "This means I conduct the basic safety training for new employees, ensure that staff members attend their annual refresher training for various programs like Blood Borne Pathogens, Injury and Illness Protection Program, and Hazard Communication Program on time. I am constantly researching and assessing new information and passing along helpful tidbits. I arrange classes on various safety issues and techniques and sometimes gives short talks, at lab-wide meetings, on specific safety issues."

Linda is also responsible for a lot of record keeping, including maintaining and updating the lab's chemical inventory and tracking material safety data sheets for each chemical. She makes sure employee-training records and equipment-maintenance records are updated and current and writes monthly status and progress reports. She is also responsible for chemical and biological waste collection and disposal.

Her preventative responsibilities include ensuring that safety systems are in place and working. She tests safety equipment (emergency showers, eyewashes, fire extinguishers, fume hoods, biosafety cabinets, first-aid kits, and so on) monthly to make sure all are in proper working order in case they are needed. She gets building repairs done through the facilities officer and serves as a liaison with the facilities manager on building problems.

Linda also troubleshoots safety issues for the staff and will call in the services of an industrial hygienist when necessary. She is the security-team coordinator, monitoring building security and related issues, including overseeing the intrusion and walk-in freezer (environmental) alarms and handling the on-call team.

Of course, Linda responds to staff concerns and any incidents and oversees staff health issues arising from accidents, which are rare. She says of her typical day: "Relentless! I never know what will happen."

Career Path

Linda had expected that after graduating from college she would conduct entomological field research on an agricultural station, but notes, "people rarely end up doing what they plan." In England, in order to obtain such a position she had to pass a civil-service exam. As part of the exam, she was required to indicate three fields in which she wanted to work. Those who passed the exam were then made available to positions in any of those areas. In Linda's case, her first interview was with a forensic lab, and she accepted a position with the Metropolitan Police Forensic Science Lab (Scotland Yard). She was hired to work in the biology department, but when she arrived for work on the first day, they were shorthanded in the chemistry department, so they asked her to help out there. Linda agreed and started examining and analyzing trace evidence using microscopy, distillations, chromatography, and computer databases.

After six weeks she was offered a transfer to the biology department, but by then she had seen what types of things the biologists handled, and what the chemists did, and decided she was much happier where she was. Linda spent all her time in the laboratory; she was not required to visit crime scenes or make court appearances. She managed the chemistry lab and became a trace evidence chemist, learning on the job.

After 14 years there, Linda followed her husband to the United States. He took a year-long research position at Berkeley, and Linda followed with a year-long leave of absence from her job but no work visa. She experienced significant culture shock, and was confronted with the fact that "there was a blue sky every single day." She recalls, "I hated that."

After about a year and a half, she went back to England for a short time but then returned to the United States. She says, "I was determined to stop thinking about things I missed and stop worrying about things I didn't like. That made all difference in being able to make a life here." Unfortunately, she was unable to find a job as a forensic chemist because of scientific conflicts between her husband and one of the prominent scientists in that field.

In 1980, she obtained a position as a research assistant in the School of Pharmacy at the University of California in San Francisco (UCSF), conducting HPLC studies on quinidines for use in clinical trials. Linda was thrilled to be back in the lab, and the science was interesting, but she still hoped to find a position closer to forensic chemistry.

Shortly after she started at UCSF, Linda got a phone call from Chevron. They needed someone with GC experience to start right away and offered a generous salary, so she accepted. Linda worked as a geotechnician, preparing oil and rock samples for GC/MS using HPLC, GC, and soxhlet extraction techniques. After several years, she took a one-year leave of absence to attend a Bible graduate study program. She returned to another Chevron company doing the same work until a repetitive-stress injury forced her to leave that position after two years.

Linda found a non-science position but was laid off after only five weeks when the company went out of business. She ended up being out of work for about 18 months, doing contract work here and there while searching for a new position. During this time, she made herself part of the forensic community—she joined the professional society, went to dinner meetings and seminars, and kept in touch with people from her first year in the United States when her husband had been working in that field.

Linda also added to her skills during this time. She took a week-long course in microscopy and found out the Sacramento Department of Justice had a scanning electron microscope that no one was using. Linda arranged to visit that lab once a week and conduct a pollen project. While there, her colleagues mentioned that they had openings and suggested she apply. She did, and it turned out the DNA laboratory needed a safety officer with a forensic background. Linda had done enough safety work at Chevron to qualify, and she accepted the position.

Linda loves using her knowledge of forensic science again and working with lots of different people. She interacts with scientists, administrators, young people, older people, contractors, and property managers and engineers. "I can use my chemistry knowledge, and I love helping other people and watching out for them."

The most difficult part of her job is that safety officers have responsibility but no authority, which is made worse when management does not support safety. Often her biggest challenge is convincing management that they should give more priority to safety concerns.

In order to do her job, Linda must be able to read the regulations, decide if they are applicable, and then apply them. She must work with all kinds of people and persuade them to do the right thing, which requires that she be a likable person. She must be organized and able to find information quickly and be able to use her chemical background to make decisions.

"Common sense is very important. A lot of basic chemical techniques and information also ends up being very important, and many biochemists, microbiologists, and DNA specialists never had that as part of their training."

Unfortunately, Linda's position is outside the regular hierarchy of state jobs, since she is a criminalist who does not do "crime work." Therefore, she does not qualify for promotion and would have to leave the DNA lab to advance professionally. She does have good benefits which have allowed her to take two leaves of absence to teach conversational English to English teachers in China. Safety officers "often go out on their own as consultants, so this career also has good post-retirement opportunities."

Advice

Linda loves her work, and urges, "Try it, you might like it!

"Networking is vital. I wouldn't have been able to do half of what I have without help from other people. Joining the American Chemical Society, and getting involved so I met people and they got to know me, was a great advantage for my career."

Predictions

Linda is optimistic about the future of the field, noting, "Universities and companies are waking up to the necessity of having someone in charge of their health and safety. This helps their audit process and their bottom line—in regards to worker's compensation and fines for OSHA violations. The field is expanding, so there are lots of opportunities."

Ruth Hathaway

Partner
Hathaway Consulting, LLC

BS, Chemistry, Huntington College, Indiana, 1979
MBA Essentials, American Business Women's Association–Kansas University, 2004

Current Position

As a partner in her own environmental consulting firm, Ruth oversees all projects involving wastewater compliance for her clients. She also oversees contracts involving either cleanup of sites or quality-assurance testing of pilot projects for various government agencies.

Like all small-business owners, Ruth spends a significant amount of her time marketing and soliciting new clients. She identifies various prospective clients "from those who want to sue others for everything, to those who want to get out from being sued." Once she has identified someone interested in her services, Ruth meets with them to discuss their needs.

Once both sides understand what the other will do, Ruth draws up a contract to make sure all the details are spelled out. She then "selects the right mix of staff to work on the project and ensure the best results for the client." She notes, "Not only scientific background and skills but also availability and personalities must match for a truly successful project."

She also manages ongoing projects, reading and approving contracts written by other staff members, conferencing with staff about the status of their projects, and offering advice.

Career Path

After a brief time in graduate school, Ruth relocated to California for personal reasons. She found a position teaching chemistry, which she had done while in graduate school. However, Ruth soon learned that she did not enjoy teaching. She says, "Chemistry had always come very easily to me. I could just look at a problem and see the answer. I had a very hard time bringing it down to the level of the students and understanding what it was they did not understand."

In 1983, another relocation brought her to Missouri, where she obtained a position at Southern Industrial Products. As head chemist, she was responsible for quantitative and qualitative analysis, sample analysis, and product safety.

She handled all the Environmental Protection Agency (EPA) paperwork, and maintained and generated material safety data sheets. A disagreement over proper procedure led her to leave the company in 1987 and start her own consulting firm. Ruth recalls, "At that time, my husband and I wanted to start a family. Consulting seemed a viable solution that would allow me to spend more time at home. The consulting business failed in less than a year, and I was back to working full-time. During that one year, however, I learned a lot. I learned I did not have a network built up, nor did I have enough experience to be of value as a consultant."

During that year, Ruth had been volunteering on the local emergency-planning committee, helping her community prepare for natural disasters. Through that work, she met people at Delta-Y Electric Company, which then hired her as a lab director. She was responsible for all (polychlorinated biphenyls) analyses and helped train linemen and supervisors in PCB handling. She was jointly responsible for corporate environmental procedures and policy enforcement.

After a few years, Ruth moved to Environmental Analysis South, where she was a quality-assurance director—a more senior position with better pay. Her responsibilities increased to include the entire chemistry section, PCBs, pesticides, herbicides, explosives, phthalates, and amino-acid analyses. She evaluated test methods and made sure the company maintained their state and federal certifications. She was responsible for the safety of all employees, provided training in HAZCOM requirements and reviewed material safety data sheets. She also periodically conducted and evaluated environmental audits and made Occupational Safety and Health Administration (OSHA) safety inspections.

During her years with Environmental Analysis South, Ruth's responsibilities continued to grow. Also during that time, she had various client company representatives suggest that she start a consulting business so they could use her services on other projects.

Ruth did some research and found out that to go into consulting, it is best to have a five-year financial safety cushion, since it can take that long to break even. After reviewing their family finances and engaging in much discussion with her husband, Ruth decided to try consulting again. She wrote her resignation letter, giving her employer three weeks' notice and allowing herself time to set up an office, print business cards, and so on.

Ruth recalls, "I planned to turn in the resignation when the lab director returned from a trip. However, my plans changed radically when the director called the day before I planned to turn in my notice and terminated me. It took me one half hour to pack up and leave.

"I also quickly learned that flexibility is a key ingredient for a consultant. The very next morning, my first day on my own, I received a phone call. A company was getting ready for an EPA audit that would occur the following day.

They had called the lab where I used to work and had been informed that I was out for the day, so they were calling me at home. They asked if I would be willing to assist them and could I be there within an hour. The answer to both was yes, of course. After the initial meeting with their corporate personnel, I was hired. When they asked for a business card, I told them I would give them some as soon as I received them from the printers. When asked for a company name, I gave the first answer that came to mind, Hathaway Consulting. The audit went well and resulted in a continuing arrangement with that company."

Within the next two months, Ruth had several other clients, all referred by state agencies. She says, "One time, I received a phone call from a Chicago lawyer. The request regarded wastewater and sampling concerns. The lawyer had called several people—whom I still do not know to this day—and every one of them had given my name as a possible person that might be able to assist him. I met with the company, and it turned out that their concerns had to do with QC/QA with the collection and analysis of their samples, one of my areas of expertise."

Ruth's business continued to grow. She hired a part-time secretary and decided to work part-time—enough to pay all her expenses (including payroll, dues, and attending national ACS meetings) and put some money away for retirement. She says, "All the clients I have are dealing with EPA problems. The understanding that I have with each is that I will not get them out of a fine but will assist them in finding a solution."

In 1999, her business started to pick up, especially during the summer. Ruth hired a chemist who wanted part-time work to assist during the busy times or when she needed a second chemist on a project. The best part of owning her own company is "being able to have high-quality scientists ask [her] to hire them—and being able to do it." She says, "This has strengthened the company and has allowed us to branch out into several satellite offices." But there have also been difficulties. According to Ruth, "The environmental field has gone through many ups and downs. Some years were difficult, not knowing if the company I was working for or the one that I owned would survive."

Ruth says that at one point she just realized that "in this field, about the only way to not to get lost in the up and down cycle of the industry is to move up the ladder and start your own company." She also liked the fact that there was low overhead. She recalls, "This was an area that I could work in without having a lab," she notes. Looking back, Ruth now says she should have branched out on her own earlier, just not quite as early as she had tried to the first time.

Ruth has a broad background in environmental regulatory compliance and strong analytical skills. By starting her own company, she was able to work to "improve the environment around us, and really see the impact" of her work. Ruth found her undergraduate research experience very valuable in preparing her for this career. "It really gave me the background that I needed. Analytical chemistry and qualitative and quantitative analysis also come in handy," Ruth advises.

Keeping up with changing regulations is crucial. "Most new regulations ... are preceded with workshops, but you have to know to look for the workshops," according to Ruth, who continues to take training to keep her knowledge base current and has averaged two to three courses per year over the last few years.

In 2002, Ruth decided to move on. She had been an occasional auditor with RGIS Inventory Specialists for many years, and they asked her to work full-time as an area manager. Ruth explains, "I basically sold the consulting company to the employees, while retaining a percentage of the ownership." After a few years, she was promoted to district manager. A recent corporate reorganization significantly changed her work environment, and she has decided to return to consulting only.

Ruth's biggest career challenge has been "trying to keep from wanting to strangle those out there who have no concept what is obtainable." She says, "I've talked to people who want to lower the amount of a pollutant simply because lower detection limits are now obtainable, but the pollutant in question requires one to drink 10 gallons of water every day for 20 years in order to have any side effects." Educating potential clients, and educating the public, is an occupational necessity.

Advice

Ruth advises, "Do not limit your focus to one area of the field—for example, don't perform only volatile organic testing. Learn ICP (Inductively Coupled Plasma), gravimetric techniques, and as much as you can. Stay current.

"While salaries in consulting are slightly higher than for those who stay in the lab, the insecurity balances that out."

Predictions

Ruth says of the future: "I think the field will continue to cycle up and down, driven by outside forces. Environmental chemistry has never been and will never be an easy ride."

Olin C. Braids

President

O. C. Braids & Associates, LLC

BA, Chemistry, University of New Hampshire, 1960
MS, Soil Chemistry, University of New Hampshire, 1963
PhD, Soil Chemistry, Ohio State University, 1966

Current Position

Olin is president of O. C. Braids & Associates, LLC, an environmental and geochemical consulting practice. He is responsible for administration and consulting work in the area of soil and groundwater contamination, contaminant monitoring, contaminant movement in soil and groundwater, and contaminant remediating. Once a contamination is discovered and quantified, treatment is required to ameliorate the threat. This treatment can take the form of physically removing the contaminant or allowing the toxic chemicals to chemically and/or microbiologically degrade. Treated water can be discharged to surface water or groundwater, and the byproducts can be disposed of according to their hazard class.

With the passage of the Resource Conservation and Recovery Act (1976), laws specific to hazardous waste disposal came into being. As a result there has been much more litigation in the environmental arena than had been the case earlier. Since many contaminants are organic chemicals, Olin's expertise in chemicals and organic matter in soil allowed him to become an expert witness.

Olin has been involved in a number of interesting cases over the years, including the Kingston Steel Drum Superfund case, the first Superfund case to go to court. This case concerned a company that recovered industrial 55-gallon steel drums, emptied and refurbished them, and then resold them to industrial clients. As regulations grew tighter, the drums were received with more and more waste content. The company bought several tanker-truck tanks and formed a lagoon to accept more waste. A couple of people then set up a subsidiary business that crushed empty drums, unfortunately leading to soil contamination in the vicinity of the crushing operation. Someone, in reviewing EPA files, found pictures showing liquid spraying out of the ends of a drum as it was being crushed, so the EPA joined the Potentially Responsible Party (PRP) group. Olin's role was to trace the chemicals in the groundwater to determine whether they followed a track that would mix with the contaminant plume from the main part of the property. He concluded that the plumes remained separate. This case took 16 months to try and had more than 2,000 exhibits. In the end, the EPA was held accountable and had to pay several thousand dollars for their cleanup mistake.

Olin is usually working on several cases simultaneously. He researches the case history and theory, and works on testimony that would counter the claims. When working on a major case, he may work on only one for a year or more.

Career Path

During the summer of his sophomore year in college, Olin was a field worker for the agronomy department, which allowed him to work outside on the experimental farm. He liked the job and decided that if he had the chance, he would combine his chemistry background with soil science and get a job in soil chemistry.

Upon graduation, Olin was offered a position with a pharmaceutical firm, but he declined it in favor of a research assistantship in the agronomy (scientific agriculture) department. This allowed him to pursue graduate studies in soil science. Olin recalls, "It was a decision that I never regretted, both for the salary and travel."

In the summer of 1962, Olin attended the American Society of Agronomy (ASA) meeting and met a professor who was a specialist in soil organic matter. The professor offered Olin an assistantship for the following year, so he changed schools and obtained his PhD. At another ASA meeting, Olin met a professor who offered him a postdoctoral fellowship working with amino sugars in soil. He then accepted a professorship at the University of Illinois, in the agronomy department.

Olin received a grant from the Sanitary District of Greater Chicago to study the application of sewage sludge (now termed biosolids) on crops. It allowed him to have "a lab technician and money to outfit the lab, which was a rare thing for an assistant professor." He recalls, "I utilized lysimeters that had been installed on the South Farm in the 1930s and which had not been used for many years."

A lysimeter is a natural soil profile that is obtained by pushing a cylinder into the soil and then excavating around it. These can be small or quite large—up to three feet in diameter and six or seven feet deep. In this case, a basement had been constructed for access, as these lysimeters were used to collect water that had penetrated the soil. The lysimeters collected the leachate from the soil, and the leachate was then chemically tested for nutrients and trace elements. The group was also able to put into place a second installation of lysimeters, constructed with drainpipes to bring the leachate to an apparatus that used tilting buckets to measure its volume, and they constructed a device that applied sludge to the plots. In this way, they could apply sludge to a plot then determine exactly what leached into the soil at specific places.

After working on the sludge project for a couple of years, Olin was offered a chance to participate in an interdisciplinary study of lead in the environment, using hydroponic plant-growth media and standard soils, to test the mobility of lead in soil and its uptake by plants.

In 1972, Olin moved back to Long Island to work with the U.S. Geological Survey (USGS). There he sampled and tested groundwater contaminated with landfill leachate from the Babylon, Islip, and Sayville landfills. They installed monitoring wells, sampled them, and sent the water samples to the United States Geological Survey laboratory in Albany. They also did field analysis for conductivity, pH, and nitrates. During his time there, Olin became acquainted with the consulting firm Geraghty & Miller, Inc. (G&M).

During the time Olin was with the USGS, Brookhaven National Laboratory (BNL), (which was originally set up to apply nuclear physics to everyday problems) was experimenting with treating septic-tank effluent by an overland-flow scheme, which resulted in reducing biological oxygen demand, suspended solids, and chemical oxygen demand. Olin was the USGS official liaison to BNL. Long Island at the time had thousands of septic tanks, which were adding to the biological-oxygen demand (BOD) and total dissolved solids (TDS) in groundwater.

TDS measures total solids (salts) remaining after the water is removed. The BOD measures the loss of dissolved oxygen over a five-day period, or some other span of time. The higher the TDS, the higher the specific conductance (electrical conductivity) of the water. For certain applications, a high TDS is necessary, and for others, it is detrimental. The higher the BOD, the higher the microbial activity. This is desirable for wastewater-treatment plants but not for potable-water systems.

In 1975, G&M was one of the few groundwater consulting firms in the United States, and they hired Olin as a geochemist to work on groundwater contamination from industrial-waste disposal. Olin says, "Having a background in organic chemistry was especially useful for the interpretation of data, as gas chromatography of the water matrix was in its infancy and precedents for the data were not available. At first, the regulators were perplexed when data came from the laboratory that indicated the presence of dissolved organic chemicals, because the reference books indicated that the chemicals were 'insoluble' in water. I gave the Nassau County, New York, Department of Health a seminar on that subject in 1976, explaining that part-per-million quantities of many chemicals could dissolve in water, although at that time they had not been very well documented." He ended up working at Geraghty & Miller, Inc., for about 25 years, except for short stints at Blasland, Bouck, & Lee, and Eder Associates.

One of the difficulties at the time was that government regulators were engineers or scientists who generally had little understanding of groundwater hydrology. Hydrology is the scientific study of the properties, distribution, and effects of water on the earth's surface, in the soil and underlying rocks, and in the atmosphere. As a result, G&M held seminars for regulators and businesses on groundwater fundamentals. Olin lectured on groundwater chemistry, data reduction, and data interpretation. At the peak, G&M held about 25 seminars a year, 4 of them intended for the general public.

While at G&M, Olin had consulting projects for many of the large chemical corporations in the United States. Monsanto, du Pont, American Cyanamid, General Electric, Pfizer, and others were investigating their processing, waste-disposal practices, and chemical-handling techniques. He was also involved with various quality assurance issues in the sampling, analytical, and reporting stages of projects.

When environmental legislation began to be passed in 1976, the lawyers got on board and filed suits against both industries and regulators. Olin became an expert witness because he understood both the chemistry of contaminants and groundwater hydrology. In 1999, he resigned from G&M and went into business on his own.

As a consultant, Olin interacts with various types of engineers, microbiologists, other chemists, hydrologists, geologists, and lawyers. He also meets nonscientists who own property with waste-disposal issues of one sort or another.

For a job like his, Olin recommends "a good knowledge of chemistry, hydrology, and the ways to apply that knowledge." He says, "I have been happy that I kept my chemistry textbooks, as many of the current problems came from activities in the 1940s and 1950s when my textbooks were new. I have been able to find references to processes and reactions that industries were using in that timeframe."

Salaries are more than competitive as one gathers seniority in this field. As a consultant with almost 40 years' experience, Olin has sometimes earned more than $200,000 per year.

He says, "What I like best about my career choice is the opportunity to do fieldwork, coupled with interpretation of the data derived from sample analysis. I also like to testify in court concerning the data I've gathered or the data gathered by other scientists. What I like least is that, as a consultant, there is no paid time off! I am thinking about retiring, so my career is close to the end, but I don't regret anything about it except that it was too short."

Advice

Olin advises, "A knowledge of chemistry fundamentals comes from chemistry courses taken in the chemistry department. Applied chemistry courses can be useful as well and are often available in other departments. Specialty information such as geology, groundwater hydrology, groundwater modeling, and so on can be taken in the junior or senior year or in graduate school. The more you learn, the better off you will be."

Predictions

Industrial accidents continue to provide work for those in this filed, according to Olin, who notes, "As much as has been done in environmental remediation over the past 35 years, there is still a way to go. This morning's newspaper had an article about an industrial plant that has yet to be cleaned up from a methylene chloride spill in 1982.

"I think there is a future in environmental work, but it is certainly different from when I started out. I was around when the Love Canal first became famous, and I was involved with it twice. I prepared an expert report on the environmental behavior of chemicals in it, and one on the public's perception of toxic substances. Now, the people who were around when the environmental movement began are in their upper 60s or older, and many are in retirement."

Dan Eustace

Health, Safety, and Environmental Manager
Polaroid Corporation

BA, Chemistry, State University of New York, Buffalo, 1970
PhD, Physical Chemistry, Brandeis University, 1974

Current Position

Dan is one of three health, safety, and environmental managers at Polaroid. As such, he has no specific assigned tasks but is responsible for helping two divisions stay in compliance with all health, safety, and environmental regulations, while also supporting corporate goals. This involves overseeing processes at three different manufacturing sites, employing about 500 people. Each of the three managers specializes in the site closest to their homes, but all three managers are on call 24 hours a day, 7 days a week. He may get several calls in one day then none for a month.

Dan says, "Each day I am surprised by whatever the current circumstances are. I start my day with a 7:30 meeting with staff, then an 8 A.M. meeting with operations, to talk about the current status, anything I don't know about yet that's happened overnight, and what we're going to do that day. I don't plan more than two days out, because I never know what's going to happen further along."

Dan does have a series of planned activities, including reporting various totals to government agencies, tracking hazardous waste materials to their proper disposal, filing interim or status reports, and so on. Dan conducts interviews with those involved if there has been an accident or incident, to explore the root cause and determine what remedial actions need to be taken. In one case, someone put an aqueous solution in a solvent-waste tank, so the waste-treatment plant won't accept it. Dan must mitigate the damages and put systems in place so the problem does not recur. A lot of his job involves knowing the people involved and the situations they run into, so he can help them solve their problems and avoid incidents without generating unnecessary work for anyone.

Dan notes, "In order to cover everything we do with only three people, we have to have an efficient system. Everything is computerized, all reports are filed online, and copies are e-mailed to the appropriate people. This way we don't waste paper and people don't have to worry about storing hard copies, but they have the information available when they need it."

Dan often uses outside consultants to develop storm water–pollution prevention plans (identifying potential sources of pollutants and outlining best management practices that minimize the potential for those pollutants to enter bodies of water) or hazardous waste–management plans. Dan says, "We have experience with all the regulations and know corporate policies and needs, but we often use consultants who are up to date on the most current rules and know the key players in the governmental agencies."

Dan also has influence at the corporate level. When a new product is brought in for development, Dan defines the risks associated with it and sets up appropriate training and practices. For example, on one project the managers wanted to use isocyanate, which reacts with water. Dan was involved in setting up a system to deliver, store, and clean up the material, in addition to setting up training for the people who were going to work with it. He also talked to the people in research, to see if the same end product could be created with something safer. He says, "We are trying to make the production process friendlier, both to people and to the environment. We are really doing green chemistry—we apply all the same principles, just don't have formal training."

Dan says, "What is most important in roles like safety, health, and environment is effective communication skills and being able to motivate others to do the right things when no one is looking. Sure, there is a regulatory role interfacing in various ways with agencies and their regulations. However, one's effectiveness on the day-to-day level is in translating the technical information to practical implementation."

He says, "In this type of work, you are never are done, and your day never ends. Eventually, you just have to decide when it's good enough. If it's within the law, always safe, practical, and makes sense, you leave it alone and move on to the next problem."

Career Path

Dan attended graduate school at Brandeis, but when it came time for him to graduate he noticed that there were many more recruiters visiting the nearby Massachusetts Institute of Technology. So, he asked the dean at MIT if he could talk to the on-campus interviewers after they were done with the MIT students. The dean approved, and Dan received three job offers from those interviews. "It never hurts to ask" is Dan's motto.

This was the time of the oil embargo and long gas lines. Dan accepted an offer from Exxon to work on a brand-new project, developing batteries for electric vehicles. "I wanted to save the world," he recalls. This was exploratory research at the most basic level. However, when Exxon found an area they wanted to explore, they invested the resources needed to make it successful. Dan started out in a group of 2 scientists, which grew to around 75 people over the next two to three years. While the group never did come up with a commercially successful product, they did invent a lot of patentable technology. Those patents were then licensed internationally, so Exxon did not lose money on the group. Toward the end of the project, Dan was involved in transferring the technology they had developed to Japanese and Austrian companies. He built prototype batteries and shipped them all over the world then debugged them when they failed.

When that project ended, Dan moved to the solar-energy department, developing solar cells. Exxon "hired and fired based on business needs, but if they saw a potential fit, they would transfer you, and give you a reasonable time to sink or swim," according to Dan. His new boss was a great mentor to him, teaching him to think about his career long term and get involved in professional societies. Dan recalls, "It was while there that I learned that no job lasts forever; you need to learn how to grow both personally and professionally. Never stop learning, and make sure you have a continuous career growth plan. The company may have one for you, but you also need to look out for your personal interests and where you want to go." Dan's boss put him on committees, where he learned teamwork, negotiating skills, how to present bad news properly, and much more. "He taught me career skills, not just technical skills," Dan recalls.

As part of their research, Dan's group worked with several smaller companies that Exxon owned, so he learned to work with outside contractors. He also traveled to universities to conduct experiments using equipment that Exxon did not own. Later on, the travel became more frequent, as he visited various locations to test out the new solar cells.

When that project ended, Dan moved "from solar cells to mud." He explains, "We were trying to devise ways to help the drilling process for oil and avoid lost-circulation problems." This involved learning about complex fluids, rheology, fluid properties and flow, and, finally, how to control flow and make it work for the process.

Dan recalls, "One of my great successes had to do with cement. As you're drilling down for oil, you send water down to cool the bit. We were looking for a shear-thickening cement—something that would fill the holes made while we were still drilling. While working on this project, I had to have a cavity filled. I asked the dentist about the filling material, which turned out to cure in about two minutes, which was just what we needed. I contacted the supplier and ordered 100 pounds, which at that time was the supply for the entire East Coast! We tested it in the lab, and it worked wonderfully. My boss was thrilled and wanted to show it to the board of directors. We did but forgot about the safety precautions, and the cement splattered all over them. Needless to say, my career with Exxon shortened considerably at that moment. At the time, I was writing papers and patents on complex fluid chemistry, but my superiors indicated that I might be happier working elsewhere and gave me nine months to find something else."

Dan used that time to do background research on the state of the scientific workplace and learned that American manufacturing was greatly in need of scientists, especially those who understood statistical tools and advanced science and technology. With this in mind, he concentrated his job search on moving out of research and development and into manufacturing. Dan attended a national American Chemical Society meeting and submitted his resume to the National Employment Clearinghouse. He had seven interviews at the meeting, four of which resulted in more in-depth onsite interviews.

One of those interviews was with Polaroid, at the time the most innovative photo company in the world, with a strong product line and new products in the pipeline. Dan was intrigued by the chemistry of making images appear instantaneously, without exposing them to light. He accepted a position as a process chemist with Polaroid and relocated his family.

As a process chemist, Dan was responsible for scaling the manufacture of emulsion products up from laboratory to pilot plant. Because emulsions are the light-sensitive components of film, all the work had to be done in complete darkness, using remote-sensing technologies. Dan initially knew nothing about these technologies but learned quickly.

In his 20 years with the company, Dan adapted to Polaroid and was always willing to take on new assignments to find a work environment that would allow him to have a good work–family balance. He turned down a promotion to project manager, which would have involved relocating again or making a 60-mile daily commute. Because of this, he did not progress as fast in his career as he otherwise might have, but he feels the tradeoff was well worth it.

Dan learned that, "You should always shoot for the highest level of performance, the best that you can possibly do in light of what you are given. You can't take a rest and glide; your career will not survive it." Companies hire and promote those who know things and who get things done. They want people who are creative problem solvers, people "who notice things and find out why." Being observant pays off. In one case, Dan noticed that in the reactor that was not working well one turbine was a slightly different color. Further examination showed that that particular turbine was made of titanium, which was reacting with the product.

During Dan's time with Polaroid, the corporation was changing. Initially, proceeds from a lawsuit allowed them to go out and buy new technologies and try new products. They had many projects that were technical successes but not commercial successes, in that they did not make money overall. Dan notes, "In many cases, successful companies are those that can give up quickly when it becomes clear that a project is not going to be a business success."

In 1995, Dan changed roles again to become a manufacturing reproducibility manager. In this position, he developed metrics and goals for batch-to-batch reproducibility of products. He identified what was important and how it should be measured, so products were on time and within specifications. He then put systems in place to ensure that the required quality level was achieved. Once these systems were in place, improvement over time was monitored.

In order to better understand how to manage variability, Dan took a three-month Six Sigma training course. Each manager entered the training with a real project that involved at least $250,000 of profit. Six Sigma involves a four-step procedure—mapping, analysis, improvement, and control. First, Dan had to define the process for his project and identify the expected outcomes. Next, he created maps that showed the inputs and outputs for each stage in the process. He then looked at the causes of variation in the process and evaluated them to find out which ones were critical. He could then focus his efforts on those aspects and reduce variability in the final product.

On his next project, Dan led an interdivisional team to develop nondestructive near-IR methods for assessing manufacturing fluids. This was part of his Six Sigma project, and he was able to replace 40% of the analytical tests in a process, and measure four or five analytes at one time nondestructively then use multivariate statistics to conduct the analysis. Eventually, the manufacturing process was running smoothly, so Dan found himself again looking for a new project.

At this time, the company was laying people off, but they needed someone to fill in for a short time as a health, safety, and environmental (HSE) manager. Dan had always been interested in environmental issues and safety, so he took the position.

This was a fairly new division for the company. Dan had to learn all the software and the filing system and learn how all the records were kept. He used his Six Sigma training to create a process map, applying what he had learned from his previous position to the new challenge.

Dan has now been in HSE for almost seven years. During that time, he has negotiated air-regulation operating permits; solved some noise-pollution problems arising from pollution-abatement equipment; implemented a site-mercury assessment, remediating, and reporting effort; and handled many other issues that arise in a chemical-manufacturing area. He has now moved completely away from experimental work. Contractors or hourly workers perform much of the hands-on work that needs to be done in this area, leaving Dan to supervise workers and handle the reporting and paperwork

Dan believes that "management is communication without using the authority of one's position, and decision making when not all the facts are known."

Since his move into HSE, Dan's company has laid off four more people, taking the department from seven people to three, while increasing monitoring from one site to three sites. Dan is starting to explore other options for his professional future. He has not found another obvious growth area, so he is exploring what interests him most. His goal is to apply what he's learned most recently to other areas. He notes, "Companies are most interested in what you've done lately."

Dan is now considering a move from HSE to molecular biology, exploring how the eye works. He has been investigating macular degeneration on his own, asking "what new learning could I apply to this field?" He explains, "Laser raman spectroscopy would let us look inside the eye in vitro and explore the chemistry as it changes over time." He attended a postgraduate workshop on Coherent Anti-Stokes Raman Spectroscopy (CARS) and is talking to one of the professors from that class about working together. Meanwhile, he is learning more about the chemistry of the eye by preparing an article for publication.

Advice

Dan advises, "Be passionate. Learn how to sell your most recent learning and skills. Find out where the opportunities are and go after them. It never hurts to ask for something, you just might get it.

"Find partners who are good at the things that you are not so good at, so the project as a whole will succeed. The team will then experience success.

"Look for the best people and emulate them. Good training is very much influenced by individuals, who help you grow and learn. Develop and maintain relationships with people throughout your career. I have been fortunate to have networks of people out there who help me when I need it, and I am able to do the same for them."

Predictions

According to Dan, "We are now working with much shorter timeframes. We need to become better communicators, to move information along faster.

"Chemistry will continue to be needed, but it will become more and more diffuse. The same skills will continue to be needed, but they will need to be applied in new ways. For example, new methods need to be developed for the analysis of nanotechnology or complex biomaterials. Traditional chemistry jobs are disappearing, at least in the United States."

ADDITIONAL RESOURCES

American Board of Industrial Hygiene (www.abih.org) aims to improve the practice and educational standards of the profession of industrial hygiene by certifying persons qualified through training and experience

American Industrial Hygiene Association (www.aiha.org) is an international association serving the needs of occupational- and environmental-health professionals practicing industrial hygiene. Their consumer brochures include several that describe industrial hygiene as a profession

American Society of Agronomy (www.agronomy.org) is dedicated to the development of agriculture enabled by science, in harmony with environmental and human values

Crop Science Society of America (www.crops.org) is a professional organization dedicated to the advancement of crop science

Division of Chemical Health and Safety, American Chemical Society (membership.acs.org/c/chas) is a membership society for safety professionals and other scientists working to promote the responsible use of hazardous chemicals

Environmental Protection Agency (www.epa.gov)

Environmental Science and Technology (www.pubs.acs.org/journals/esthag) is a journal published by the American Chemical Society

Laboratory Safety Institute (www.labsafety.org) is a nonprofit, international educational organization for health, safety, and environmental affairs

National Research Council. 1995. *Prudent Practices in the Laboratory: Handling and Disposal of Chemicals.* Free online edition at www.nap.edu/catalog/4911.html.

Occupational Safety and Health Administration, United States Department of Labor, (www.osha.gov) publishes a biweekly e-news memo

Plog, Barbara A., Quinlan, Patricia J., eds. 2001. *Fundamentals of Industrial Hygiene*, 5th ed. National Safety Council: Itasca, IL.

Soil Science Society of America (www.soils.org) aims to advance the discipline and practice of soil science

Union of Concerned Scientists (www.ucsusa.org) is a group of citizens and scientists seeking solutions to environmental problems

Watershed Management Council (www.watershed.org) is a nonprofit group of professionals and others interested in the proper management of watersheds

9 Chemistry and People

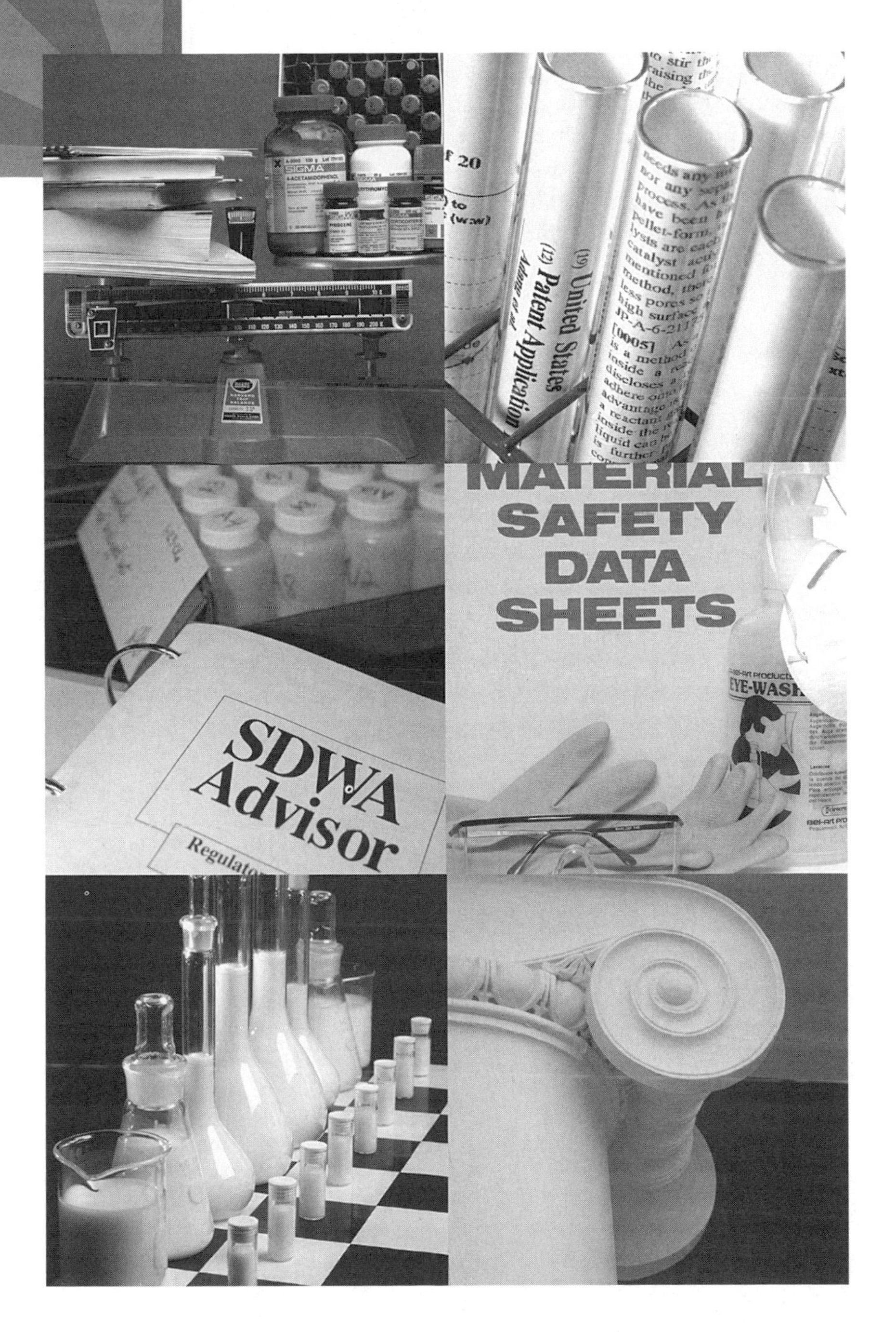

For those who really enjoy working with people and helping others achieve their personal and professional goals, human resources (HR) can be a rewarding profession. Not only do large and small companies need people to manage personnel matters, but placement agencies, headhunters, recruiting firms, and others need people who understand science, can translate job requirements into skill sets, and can find the people with those skills. Within a company, HR professionals help employees identify and achieve their career goals, and match those goals to corporate goals.

With the shifting market forces resulting in more part-time, temporary-to-hire, and contract positions, the ability to find the right people and match them with the right positions is a valuable skill more and more in demand.

* * *

In any business, the people involved are what make it work. Find the right people, and things move smoothly and the business grows. With the wrong people, or the lack of the right people, the work process is impeded, and the whole company suffers. Finding people with not only the right skills for the job but also the right personality, motivation, and all other factors to fit the needs and personality of a particular company at a particular time is like working a jigsaw puzzle—without any guarantee that all the pieces actually exist.

As the workforce becomes increasingly mobile and professionals routinely move from job to job and city to city, being able to match people and positions in an ever-shifting marketplace is a challenge that many enjoy and find incredibly rewarding. Within a large company, human resources professionals may specialize in hiring, firing, benefits, particular types of positions, or some other aspect of the job. Most medium-to-large companies that hire chemists (including pharmaceutical companies) have someone whose job is technical recruiting. There has been some outsourcing of this function, but this does not appear to be a trend.

In addition to company's employing human resources professionals, many scientific staffing agencies also employ staff to identify and screen candidates for companies. In some cases, the staffing agencies identify candidates who are then hired outright by the client company (placement agencies); in other cases, scientists become employees of a staffing firm and are then "contracted out" to the client company. Increasingly, staffing companies are offering insurance and savings plans so that people working as contractors for extended periods of time can have almost identical benefits to those in traditional permanent employment.

Contract work is on the rise, and not just for clerical workers. A growing number of scientific professionals at all degree levels are taking part-time or temporary positions as part of their long-term career strategy. For some people, it is a way to try out a variety of different companies and positions, to find what works best for them. It can also be a way to ease into, or out of, the workforce when personal circumstances change. For employers, using temporary or contract workers allows them to quickly ramp up, or down, in reaction to rapidly changing market conditions, and to address employee absences, temporary skill shortages, seasonal workloads, and special projects. The staffing agency is an advocate to help solve short-term problems.

Some companies are now also using contract-to-hire assignments, which basically lets them try out a candidate for a specific amount of time before committing to permanent employment.

Working in human resources requires a lot of people skills—reading people; negotiating; matching potential employees' skills, abilities, and desires with corporate needs; and so on. It requires attention to detail in managing benefits and making sure contracts are accurate and fair, and auditing records to ensure compliance with fair-hiring practices and the Sarbanes–Oxley Act.

Profiles

Trish Maxson

HR Group Director, Coatings Businesses
Rohm and Haas Company

BS, Chemistry, Michigan State University, 1981
PhD, Physical Chemistry, University of California, Berkeley, 1988

Current Position

Trish Maxson is the human resources group director for the coatings business of Rohm and Haas, a $7–billion specialty chemicals company. As such, she provides oversight to the business human resources (HR) support for their three coatings businesses and acts as the global strategic HR business partner for the largest of the coatings businesses. These units together include more than 4,000 people and account for more than $3 billion in sales.

A typical day for Trish is full of "meetings, meetings, and more meetings." She says, "Because the company is restructuring HR, including outsourcing transactional tasks, I currently spend about half my time defining the new organization, spelling out roles and competencies for HR employees, and providing guidance for HR processes such as compensation structure and talent assessment. The other half of my time is spent with the leaders of the businesses

I support, providing both individual coaching and workforce planning, organizational design, and talent assessment and development support."

For example, Trish meets with an outside vendor who has been hired to evaluate the grade level of all jobs within the organization and provides details on the jobs and calibration for their scope. She meets with senior leaders in the company to assess the potential of specific people in their organizations, with the ultimate goal of providing the right development opportunities for those employees so they can achieve their professional potential. These meetings require a fair amount of preparation on her part, to make sure she has all the information needed on the specific jobs and people under discussion.

Trish also manages a small group of other human resources professionals, and she spends time helping them support their clients.

She notes, "In the cracks of my day, I spend time with people, hearing their issues, fears, triumphs, concerns, and so on."

Career Path

Trish went to Rohm and Haas straight out of graduate school, as the result of an on-campus interview. She started out as a research scientist then became a technical service manager.

After she'd been at Rohm and Haas a few years, Trish had an unexpected opportunity. The company decided to train scientists in the organizational design processesto help upgrade the organization's ability to work in teams, improve project management, and so on. Trish spent a year as a facilitator, helping people become more effective in their work. This gave her a first taste of what it could be like to really pay attention to interpersonal dynamics.

Shortly thereafter, she was asked to be on a team charged with finding the next technology platform for the company. She and about nine other "really smart, driven people" spent two years researching a huge number of technologies, companies, and markets, to arrive at a recommendation for a growth plan for the company.

Ultimately, the company didn't follow her team's advice, and almost simultaneously they shut down a new small business where Trish was a product manager. She recalls, "I felt burned out by what felt like pushing a rock up a hill over and over and always being unsuccessful at keeping it there."

Trish had earlier realized that she was happiest when "listening to people and helping them solve their problems, rather than listening to chemistry and solving lab problems." She decided to would pursue a PhD in psychology and transition to a more people-oriented career. She switched to the HR department because it was a somewhat related field and allowed her to continue to earn an income while going to school. She was initially hired for a three-month temporary position, which became five months, then eventually permanent.

Trish didn't intend to have a career in HR. She muses, "I thought of it as a job to have while I retooled myself. Through a combination of good luck, my own particular set of skills, combined with everything I've learned in the psychology program, I've had an astounding career in HR. I recently quit the psychology program after receiving a master's degree because I am content in this current career path—for now."

Trish interacts with HR people and line managers around the world. She still has very close ties to research, because of her background. The best part of her jobs is that she is able to help people in a fundamental way, which often involves helping them gain insight into themselves, and is able to effect organizational change. She says that the worst part is "dealing with all the HR minutia. . . . all the necessary transactional stuff that gets people paid, makes sure their benefits are in order, tracks their vacation days, and so on."

The Sarbanes–Oxley Act of 2002 has affected how she does her job. The act requires strict controls and audits on raising pay, adding people to payrolls, and so on. She explains, "Imagine a business unit of 16,000 people sending a monthly report of new hires and personnel changes, all of which must be tracked down and checked." Trish spends some of her time doing these tasks and auditing others, but the more important job is working to implement automatic procedures to make the process run more smoothly and require less manual data checking.

Trish gets to work across the whole company, interact with people from all over the world, and influence the direction of the organization to an extent not many jobs, other than managing director, offer. The challenge is that her role is influence, not authority, which can be frustrating at times. She has input into what people do and what direction the company takes, but the final decision is often in the hands of other people.

In order to work in HR, Trish feels it's important to "understand organizational design, have some understanding of what makes people tick, have analytical skills, a tolerance for endless meetings, and a willingness to be extremely tactful and tactical at times."

Trish finds that with her chemistry background she gains "instant credibility . . . especially in the research community, which is not always easy for HR people." She says, "I have a fundamental knowledge of our products that helps me understand how we make money, which then has an impact in the people part of the organization. Knowing how statistics work, being very analytical and unafraid of IT tools—spreadsheets, and so on—all started because of my technical background."

"There are so few chemists that turn to HR as a career that it is hard to generalize about salaries, but they are probably roughly comparable to those of laboratory chemists if you take number of years of education into account.

In my own case, I believe I am way ahead in salary from where I would be had I stayed in a more traditional career."

As for her future, Trish says, "The next job in the company is the vice president of human resources. I don't know that I would be the successful candidate for it, but I do think that, if I want to, I could find a position as head of HR for a technology-based company. I could also probably start a second career as an executive coach."

Advice

Trish advises, "Make sure you have a fundamental interest in people—what makes them tick, a desire to help."

Predictions

"Human resources will always be necessary," Trish observes. "The trend is for the polarization of strategy and tactics, with tactics moving to outsourcing providers." This means the company employees will decide what should be done, but the details and implementation are contracted out to other firms. "The roles that remain within the organization require smarter, more-strategic thinkers than in the past. Some people have predicted —and would like to think—that we will someday have chief psychological officers. I'm not sure I agree with that, but there is a big demand for executive coaching that touches HR."

Paul Erskine

Account Manager
Aerotek Scientific

BS, Chemistry, University of Loughborough, Leicestershire, United Kingdom, 1996

Current Position

Paul Erskine is an account manager for Aerotek, a national recruiting agency that serves a wide variety of industries. Paul's specific job is to sell technical-staffing services to companies that have openings for people with scientific backgrounds. He convinces them that in the long run it is more cost-effective to partner with his company and let Aerotek find and screen both employees and long- and short-term contractors.

His main responsibilities are maintaining and servicing current accounts and leading and developing a team of technical recruiters. The technical recruiters are the ones who network with job seekers and find the right people to fill the positions at the companies that have contracted with Paul.

Paul's typical day varies from "busy to extremely busy." He says, "There is never a boring moment or day. No two days are the same." Paul identifies new companies that might benefit from his services and contacts the decision makers at those companies. If he can get a referral he makes a "warm call," otherwise he makes cold calls to see if they are interested in his services. If they are interested in learning more, he may have an in-person sales call at the client site. If it is a person or company he has worked with for a while, he may have a business lunch or business happy hour to touch base and make sure he understands their current needs.

Paul conducts daily recruiter meetings to prioritize the work and to decide which recruiter will work with which company and to fill which specific positions. He also has to prioritize—some clients and positions need more urgent attention than others.

As is typical of most HR jobs, Paul's job requires that he spends a significant amount of time writing reports, to track the status of each account.

Career Path

After college, a friend of a friend of Paul's helped him obtain a position as a lab chemist at the Feed Oil Company in Liverpool, England. There he was responsible for the quality control of raw materials, both in process and finished products. The finished product was blended feed oils for cattle and poultry. Paul performed fatty-acid composition tests and PCB–pesticide ratios using GCs and some wet-chemistry techniques. He enjoyed managing the blending process, working on new products and blends and special projects that let him help clients with problems. However, after a while, the job got repetitive, and Paul decided to look for something more challenging.

He wanted a challenging but rewarding career and thought corporate recruiting might fit the bill. He talked to a friend who worked at Aerotek, leaned about the company, and liked both the company culture and the people with whom he would be working.

While most people don't move from a lab into sales, it worked for Paul. He says, "One of the most important things you need to work in this field is a knowledge of skill sets and the industry that you are serving; and having spent some time working in the laboratory means I really understand what my clients are looking for. Understanding what the client needs and what type of person they are looking for, and then being able to explain that to non-technically minded people—recruiters and HR managers—is vital."

Paul started as a recruiter, finding people to fill positions defined by his manager. At Aerotek, everyone starts as a recruiter "to make sure they receive the right training and experience—recruiter and sales training." He learned how to make cold calls, and evaluate candidate's knowledge, skills, and abilities to match them with clients' openings.

After a few years, he moved up to account manager, responsible for finding the companies doing the hiring and for keeping track of the positions needing to be filled.

In many ways, Paul feels he is running his own business. He explains, "I am given a sales territory with definite boundaries, I make cold calls, set up meetings, do presentations about my company's services, negotiate terms and conditions for contracts, and come up with creative ways to help and service our clients. I also manage the process of finding qualified candidates for them, as well as the interview process and successful placement of the people they are looking for. I manage several recruiters, giving them daily direction, so they can find and screen candidates. I help my recruiters develop and fine-tune their skills and help them get to the next step in their careers. My team—recruiters, a customer service associate, and me—treat it like it is our own business, and this is why we enjoy success and the success of our customers."

Paul works with a wide variety of people, and knowing what they require is paramount. He must be able to ask the right questions to find out what they really need. He says, "You have to like people and enjoy getting to know people. People can change their minds, so you have to be able to take a hit on the chin and keep going." Paul gets great personal satisfaction from finding someone that perfect job for which they have been searching.

The salary potential is far more than for traditional lab positions. While the base salary is comparable, commissions can take salaries much higher. Paul observes, "The financial compensation is there if you work hard and do the right things for your clients."

The next obvious step in Paul's career would be sales manager, managing a team of account managers. The next step from there would be director, managing and guiding an entire office or offices, including sales managers, account managers, recruiters, and customer service associates.

Advice

Paul suggests, "Network with people, attend breakfast, happy hours, and so on. It is all about who you know! The more people you know, and have real professional relationships with, the better you will do in this industry."

Predictions

Paul sees the field expanding, saying, "It is still very much a growing industry as more and more companies find it hard to get the qualified employees they are looking for, due to the so-called war for talent. Also, temporary help or temp-to-full-time situations are becoming more and more common as companies are not allowing people to give references on past or current employees."

Joel Shulman

Manager of External Relations and Associate Director, Corporate Research (Retired)
The Procter & Gamble Company

BS, Chemistry, George Washington University, 1965
MA, Chemistry, Harvard University, 1967
PhD, Organic Chemistry, Harvard University, 1970

Current Position

Joel Shulman spent much of his career at Procter & Gamble (P&G) as a corporate recruiter, managing the hiring of PhDs, medical doctors, dentists, pharmacists, and veterinarians. He not only established hiring standards but managed the recruiters who visited the universities, and developed strategies for hiring specialized scientists when needed. As manager of doctoral recruiting and university relations, he explains, "In a nutshell, my job was to find the best candidates, make sure these candidates knew about opportunities at P&G, and work with hiring managers to make sure we hired the best."

In a typical year, P&G conducts interviews at about 30–35 schools (mostly departments of chemistry), with some 40–50 recruiters making visits to these schools. Joel would select the schools that would be on the recruiting list, select and train the interviewers who made annual trips to the schools, and debrief recruiters when they returned from interviewing. His office would set up all recruiting visits and maintained year-round contact with the schools. Each year, he would invite some of the university contacts to visit P&G in Cincinnati, to learn more about what they do and how they use chemists. To cultivate goodwill with the faculty, Joel also managed graduate fellowships that were given to top students by P&G, either through the American Chemical Society or through individual chemistry departments.

Proctor & Gamble would normally receive up to 5,000 applications a year from doctoral-level applicants (including perhaps 500 people interviewed on campuses). It was Joel's job to review all these and determine the appropriate response—sending a rejection letter, requesting more information, sending the file to the appropriate hiring manager(s), or sometimes phoning or e-mailing references. If more than one hiring manager was interested in a candidate, Joel would determine who would issue the invitation for an interview.

In addition to campus interviewing, Joel managed the sourcing of candidates for openings. For nonstandard positions (e.g., doctors, dentists, veterinarians, and pharmacists), he would work with the hiring manager to find out exactly what was needed and then develop a strategy to find candidates. He explains, "This might be running an ad or enlisting the services of a search firm—head hunter, both of which activities were handled by my office. When you wanted a PhD epidemiologist or pharmacoeconomist, finding candidates could be very challenging."

Joel's job also involved setting starting salaries for all new doctorates, from newly minted PhDs to experienced MD and PhDs, and determining when signing bonuses were appropriate to attract top candidates. He also set strategies to find top underrepresented minority candidates, including running a three-day symposium in Cincinnati called "Research and Technical Careers in Industry," an event to recruit minority scientists.

Since about 10–15% of the doctoral-level hires were foreign nationals, Joel coordinated the process to bring these scientists into the company, working with outside attorneys to obtain appropriate work visas and setting strategies to sponsor foreign nationals for permanent residency (green cards). He says, "This multiyear process was a time-consuming one for me and my office."

Upon his retirement in 2001, Joel decided to give back to the chemistry community, and he found a position where he could add to the graduate school experience what he, as an employer, had found was missing from the student's education.

Joel is now an adjunct professor of chemistry at the University of Cincinnati, working 40%-time, for which he is paid at about 40% of a full professor's salary. In reality, he works close to full-time when he is in town, but he is free to take extended time off so that he can act "somewhat like a real retiree." However, as his wife describes it, "Joel has retired from his salary but not from working." Joel prefers to describe his position as "similar to going back to college, but without having to take courses—or exams" and likes that he can pick the "extracurricular activities" in which he's interested.

As an adjunct faculty member, Joel's responsibilities are threefold. He works with graduate students to educate them about career options and prepare them for the job market, he runs the Department of Chemistry's Industrial Affiliates Program, and he teaches some courses in organic chemistry and participates in faculty service.

Career Path

As do all students nearing the completion of their PhD in chemistry, Joel had a major decision to make in 1969–70: Did he want to seek an academic or an industrial job? Also, like most students in this situation, he had very little knowledge of what research and development in industry was like. Nonetheless, for a variety of reasons, he chose to go the industry route. He recalls, "Not the least of these reasons was my perception that my time would be less my own if I had to fight the tenure battle in academia. This was important to me because I had a young family as I began my working career, and I wanted the luxury of spending time with my children as they grew up."

Thus, upon graduation, Joel accepted a job with Procter & Gamble in Cincinnati. One reason he chose to take a job with P&G was the opportunity to be involved in a wide variety of activities beyond bench research. This proved to be the case during his first 23 years at P&G. Starting with bench research and moving soon after into management roles, Joel had assignments in virtually all of the company's product areas: foods, coffee, paper, fabric care, beauty care, over-the-counter drugs, and prescription drugs. He also became active in PhD recruiting early in his career, making annual recruiting trips to Harvard, Massachusetts Institute of Technology, Yale, and Princeton, and hiring dozens of PhD's to work in the organizations for which he had responsibility. He says, "I was also able to satisfy my lingering curiosity about academic employment by teaching graduate organic chemistry courses at night for three years at Xavier University in Cincinnati during my early days at P&G."

In 1993, Joel was appointed manager of doctoral recruiting and university relations for P&G. He says, "This was a major career change for me, since it removed me from line management of research and put me into a staff role in which I supported all areas of the company. Primary considerations in my selection for this position were my broad experience with all the business areas of the company and, of course, my long-standing interest in recruiting." In his new position, Joel coordinated the recruiting activities of some 50 PhD recruiters who visited about 35 universities each year; established hiring standards for doctoral hires (including not only PhD's but also medical doctors, dentists, pharmacists, and veterinarians); and developed strategies for finding and hiring highly specialized scientists when needed. In his first year in this role, P&G hired 117 doctoral-level scientists (a record high that still stands) In a typical year; they hire 30 to 50 doctoral-level scientists.

In 1996, Joel's responsibilities were expanded to include external research grants, interactions with government laboratories, and technology acquisition from Russia and China. He says, "I viewed my overall role as being charged to bring new technical capabilities into the company. Ancillary to my role within P&G, I became the company representative to the Council on Chemical Research, the Government–University–Industry Research Roundtable, and the American Chemical Society, where I was appointed to the Committee on Corporation Associates and the Career Services Advisory Board."

Joel's experiences with doctoral recruiting and university relations made him very aware that PhD programs in the sciences do an excellent job of training students to conduct independent research and to enter a position in academia, but they miss out on the opportunity to prepare them for a career in industry. He notes, "This is a major outage for PhD chemists, 65% of whom go into industrial positions. Much has been written about deficiencies in new PhDs going into industry—particularly, less-than-desired communication skills, ability to work as part of multidisciplinary teams, and knowledge of what is important in industrial versus academic research. All these anecdotal outages were reinforced by my observations of new PhDs at P&G."

As Joel began to contemplate retirement, he thought about these outages in new PhDs and how he might help improve the situation. This led to several discussions with Professor Marshall Wilson, head of the Department of Chemistry at the University of Cincinnati, who offered him an adjunct professor position when he retired from P&G in 2001.

The first thing Joel did upon starting his new position was to initiate (with Anna Gudmundsdottir, an associate professor in the department) a course for third- and fourth-year graduate students called Life after Graduate School. The course is divided into three parts—the real world, skills needed, and how to find a job.

The first segment addresses the question "What do chemists do in the 'real world' and what kind of opportunities are there beyond bench research?" Joel says, "As I knew from my personal experience, newly minted PhD chemists have very little knowledge of careers in chemistry beyond what they see their research advisors doing. Thus, in the course, we talk about day-to-day activities of chemists who teach at non-research schools. We cover what it is like to work at large and small companies, in government labs, in tangential careers such as products research, regulatory science, and patent law."

The second part of the course covers the skills that are necessary to survive in the real world. Joel points out, "Communication skills, of course, are vital to any professional career. Technical writing and technical oral communications are, hopefully, taught to graduate students. However, we cover areas like persuasive writing, and communicating with people not trained in your area, or not technically trained at all. We talk about intellectual property issues and ethics in research. We allow students to participate in team problem-solving exercises."

The final section of the course deals with that all-important issue of how to find the job that is right for you. Joel explains, "This section draws heavily on the workshops provided by the American Chemical Society's Department of

Career Services. We talk about the postdoctoral experience: Do you need one to get where you want to go? What should you look for in a postdoctoral position? What about an industrial postdoctoral position? We cover in depth putting together a cover letter, a resume, or CV, and a teaching philosophy—if you want to find a teaching job. We talk about interviewing skills and, most importantly, do an extended—up to two hours—mock interview with immediate feedback with each student. Beyond the course, I work on an ongoing basis with graduate students as they conduct their job searches."

In addition to educating graduate students about the real world, Joel also runs the department's industrial affiliates group. This group had been around for about 20 years, but prior to 2001, it consisted of about seven companies in the Cincinnati area that contributed money to the department, which was used to supplement the teaching stipends of graduate students. Companies received a few perks for their contributions but gave money mainly to be good corporate citizens. The Industrial Affiliates Program was run ad hoc by the department head. Joel's industrial background let him see that "the University of Cincinnati Industrial Affiliates Program was run by academics, for academics, with little knowledge of what companies might want in return; and, there were models at other universities, aspects of which could profitably be incorporated into the University of Cincinnati program." Joel formalized the Industrial Affiliates Program, including adding aspects attractive to companies. As a result, the program now has 14 members, and the funds available to supplement student stipends have nearly doubled.

The final aspect of Joel's position, and the one he says has made him "feel like a real member of the faculty" is the opportunity to teach organic chemistry. He explains, "I have taught a graduate course in organic synthesis, and some sophomore organic chemistry. In the latter, lecturing to 130 students is more of a challenge than I had imagined, but it has also given me the opportunity to introduce a few innovations. One of these is to break out for the first time in the course an honors recitation section for the top 10% of students, taught by me rather than by a teaching assistant. This provides a challenge both to me and to the students, with new material tailored to the interests of the students being covered each week."

Despite the fact that Joel is part-time, he is treated as a colleague by the faculty. He says, "I participate in faculty meetings. Some would say that is a negative rather than a positive! And I serve on several departmental committees, including Safety Research Experiences for Undergraduates, and Graduate Recruiting and Admissions—which takes advantage of my P&G experiences."

Thanks to his recruiting background, Joel has recently been asked to lead the recruiting of graduate students for the Department of Chemistry. This involves coordinating faculty visits to feeder schools and putting together materials to send to students, among other things. He also serves on the department's long-range planning committee, where he brings valuable, practical experience to academic colleagues who have never been through this process. Finally, Joel is on the department's undergraduate curriculum committee and is in the process of developing a formal program of industrial internships for their undergraduate chemistry majors.

Advice

Joel notes, "A major lesson from my experience with career changes is that, with a little planning, you can set yourself up to move in directions that interest you. For example, by demonstrating an interest in doctoral recruiting early in my career and gaining experience with a wide variety of company divisions—and working with a wide variety of people, I was in an ideal position to become manager of doctoral recruiting at P&G. In fact, I had declined to be a candidate for this position twice earlier in my career at P&G because I did not feel at those times that I was ready for it. And, of course, all my experiences at P&G provided me with the background to add value to the Department of Chemistry at the University of Cincinnati, offering a perspective not generally found among university faculty.

"Graduate education in chemistry would be enhanced if more industrial chemists considered a 'second career' in academics to help educate students on what life after graduate school is really like."

Predictions

"Overall, I think PhD chemist hiring by larger companies will remain rather static in the foreseeable future," Joel says, "with slightly increasing opportunity for jobs as a PhD in service companies—analytical services, especially—and small companies. The only change I see in graduate education is an increasing emphasis on collaborative research. The National Science Foundation is emphasizing multi-investigator research projects in their funding, which will translate to more team approaches to projects—something industry would also like to see more of."

ADDITIONAL RESOURCES

Aerotek Staffing Agency (www.aerotek.com)—technical staffing agency

American Staffing Association (www.americanstaffing.net) is an association of staffing companies

Industrial Research Institute (www.iriinc.org) is a business association of leaders in research and development

Kelley Scientific Services (www.kellyscientific.com) provides scientific staffing to a broad spectrum of industries including the chemical, cosmetics, food-science pharmaceutical, biomedical, consumer-products, environmental, medial, clinical, petrochemical, and clinical-research industries

National Association of Personnel Services (www.recruitinglife.com) is a recruiting and staffing community

Society for Human Resource Management (www.shrm.org) is an international professional organization with over 500 local chapters in the United States

Yoh Scientific Services (www.yohscientific.com) provides long- and short-term temporary and direct placement of technology and professional personnel

10 Chemistry and Computers

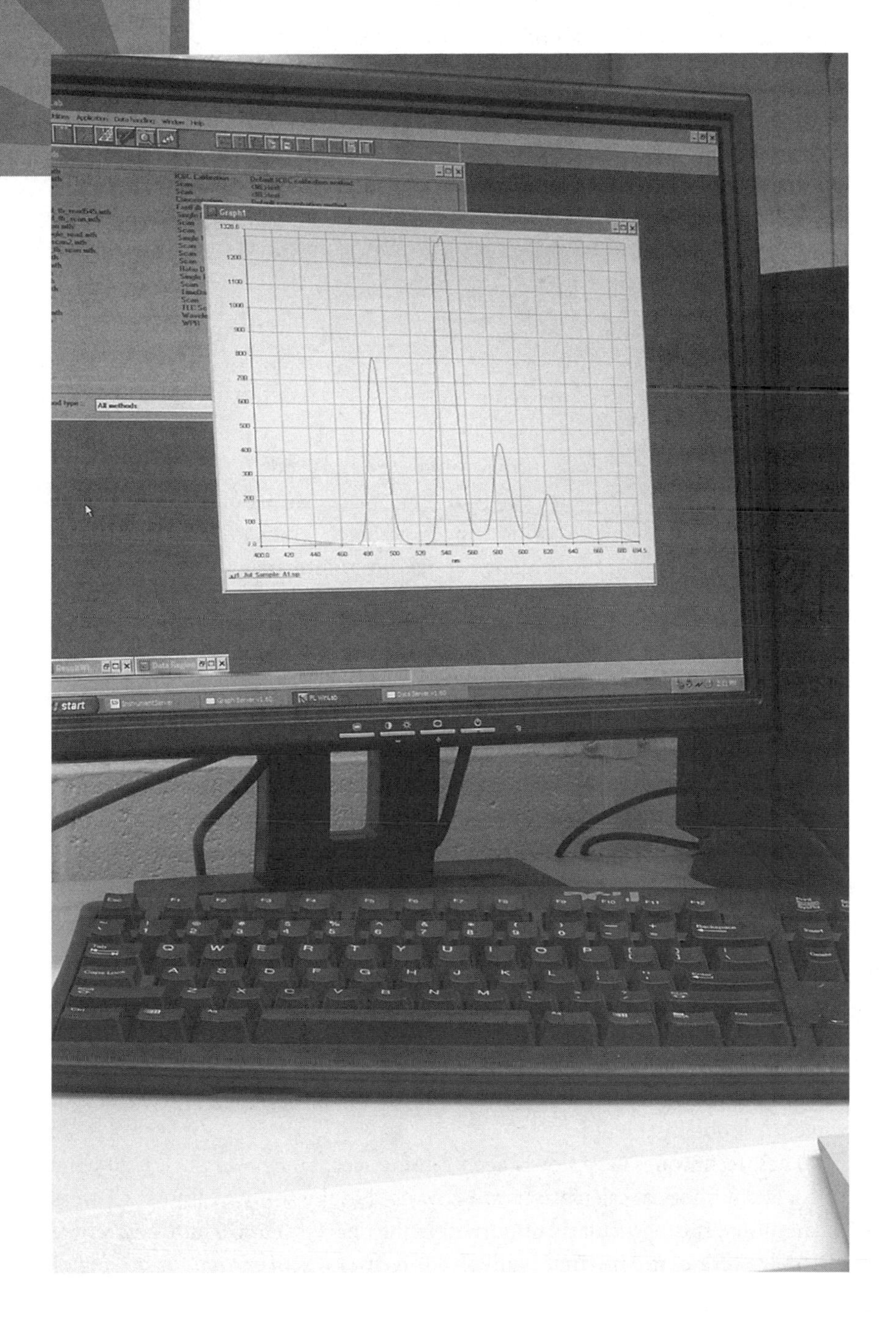

IT IS PROBABLY not an understatement to say that virtually every job involves computers in some way. In many cases, computers are simply tools, used to facilitate a traditional task, such as writing a report or balancing the books. But, in some cases, the computer itself becomes the focus and allows new career paths that previously did not exist. These include scientific programming and using sophisticated algorithms to predict properties and structures, or using databases to manage huge amounts of chemical information.

* * *

VIRTUALLY everyone uses computers in their work to some extent. However, there are those for whom the intricacies of algorithms and programming form the major part of their working lives. If those who develop the algorithms and interfaces do a good job, both in the underlying science and in the implementation thereof, there is another whole class of scientists who use that software product to make predictions, provide analyses, and guide future work.

The days of the computer programmer or user sitting alone in a cubicle, calculating numbers and throwing answers across the transom for someone else to take the next step are long gone. Today programmers and computational scientists work as part of a team, collaborating with those who use the results of their work.

Computational chemistry has become a recognized specialty in its own right, and a number of institutions now offer degrees in this field, which includes quantitative structure activity relationship (QSAR) (the use of statistical methods to predict the activity of new compounds based on the activity of previously tested compounds), molecular modeling (the use of computers and specially designed software to predict the properties of compounds based on a three-dimensional shape and its characteristics), computer-aided drug design (CADD), docking small molecules into proteins and scoring the "goodness of fit," combinatorial library design, protein three-dimensional structure prediction and homology modeling, molecular similarity measurements and searches, and many other techniques. Chemists working in this field may specialize in a certain technique, but most are more interested in ways to solve particular problems. They want to help decide which compounds to make next or which ones to take to the next level of testing and will use whatever methods work best to solve the problem at hand.

These techniques have slowly been gaining acceptance over the last 20 years or so, and numerical calculations are now used in the study of almost all areas of chemistry. It is particularly important in the pharmaceutical industry, where virtually every company that is involved in drug discovery has some sort of

CADD effort now, and the materials-science field, which includes polymers and biomedical material properties. Some companies use commercially available software, some develop and write it all in-house, and some use a combination of existing and in-house software.

Pharmaceutical companies are one of the biggest users of computational chemistry, both in the prediction of new active structures, and in the prediction of absorption, distribution, metabolism, excretion (ADME), or toxicity properties of structures. Medicinal chemists always have more ideas for structures than time in which to make them, and computational chemistry is often used to help rank which molecules will make the best drug candidates. This requires close collaboration between the computational chemists, to balance predicted activity with synthetic feasibility. Since a single computational chemist typically supports more than one medicinal chemistry project, prioritizing and time-management skills are also important.

In addition to predicting structures and properties, software must be able to draw chemical structures, assign scientific names, and much more.

Even for those who are primarily users rather than developers of the software, some background in programming is helpful. Knowing how to write a program ensures a familiarity with computers and enables one to attack problems oneself by modifying the existing code to work in a new way.

Computational chemistry is not currently one of the more visible chemistry professions, but this is slowly changing. It is also excellent preparation for careers in weather forecasting, meteorology, financial modeling, and bioinformatics.

Profiles

Joseph M. Leonard

Principal Scientist
Abbott Bioresearch Center, Abbott Laboratories

BA, Chemistry, University of Pennsylvania, 1978
PhD, Chemistry, Duke University, 1982

Current Position

As a principal scientist, Joe fills several roles. He was originally hired to design and implement an in-house molecular-modeling system, which would let the company implement its own "new science" and make them less dependent on commercial software and vendors. Joe explains, "We're not the first group to attempt to develop in-house software, but we're trying to make it professional-caliber and thus somewhat author-independent. We're trying to use developer time as efficiently as possible and so are trying to work in an interpreted, high-level language—Python. We're trying to present a single environment for both development and user scripting, which reduces maintenance overhead. To date, we've been successful.

"Now that the tools are in place, people can come in with a journal article and ask 'Can we do this?' As long as the methodology is relatively well described, we usually can provide something for the modelers or medicinal chemists to work with. We've been able to reproduce several methods described in the literature, with accurate results, and then extend them and apply them to our own projects."

Joe describes his typical week. He says, "There's debugging—always more than I'd like to see. This week, I've been implementing some new molecular descriptors, which have required the implementation in software of a new method for estimated LogP." Once code gets debugged, it has to pass the test suite, and new tests have to be written for anything created from scratch. He explains, "It's these tests that give us confidence that we can keep this stuff working, and port it from Linux to SGI—or even to Mac, to keep it usable as hardware requirements change. We assume we could port the non-graphics to the PC without too much difficulty, but our environment does not require this today."

Joe continually asks the modelers whether they have all they need to do their work. As long as they do, he can concentrate on integration and support. "I don't think they've got what they want," he says, "but until they're better able to articulate their needs, or we see some 'new, exciting news' at conferences, we're in a bit of a lull. We can't do protein–ligand scoring correctly and ADME and toxicity prediction is data-limited."

Joe believes that the key ability needed for his job is problem solving—the ability to turn an idea into something others can use and to debug it when it breaks. He explains, "I think it's important to have a strong technical basis in the field when doing applications work, so I can act as both designer and developer. I know what the users want and need, because I've been them. I've tried to keep abreast of the literature, and especially tried to watch what's of interest and what's not."

Joe is often asked to provide advice on which techniques and products might be applicable to specific modeling project requirements. He participates in reviews prior to purchase of hardware and software, investigating the requirements of both the project and the company, assessing the available options, and making recommendations. If there's any time left after this, he does "a little QSAR modeling."

Joe says, "Google has been an amazing resource for this work, since there's usually something somewhere talking about a paper or methodology. Our ability to work would be noticeably reduced if I was limited to traditional libraries and journals. The computational chemistry e-mail list—ccl.net—is also a good resource, as long as one doesn't overuse it."

Career Path

Joe began working with computers in 1971 when his high school got a PDP-8 (with 4K of memory!). In college, he worked with a few others on a chess program and even did a year of football scouting work. He says, "One did what one had to get to time on the computer—the machines were a little bigger as you recall."

Joe enjoyed both chemistry and computers and wanted to combine the two into a career. Theory seemed the only way to do that when he graduated from college in 1978, so he went to graduate school in chemistry and joined a quantum-mechanical group that developed its own methods. He recalls, "I knew how to program back when it was somewhat rare, which made me desirable in theoretical groups.

"Looking back, I wish I had known about the early modeling groups that existed in 1982, but there just wasn't a way to get the information. What theoretical connections our group had didn't cover enough ground, and things like CCL were way off in the future. People graduating even four or five years later had 'computational chemistry' starting to make sense as a career, and postdoctoral and industrial positions were available."

When Joe earned his PhD in 1982, the job market was tight, but he obtained a postdoctoral position at EG&G in Idaho Falls, where he researched and implemented application codes to support research projects, including gamma and X-ray spectrometers.

After a couple years, Joe wanted to get a little closer to "mainstream chemistry," and through a former colleague he obtained a position with the U.S. Army at the Aberdeen Proving Ground in Maryland. During his four years there, Joe did a little bit of everything, including developing and maintaining a molecular modeling, analysis, and display system, training users, developing documentation, purchasing new hardware and software, and conducting some theoretical research to predict the biological activities of candidate compounds. Joe welcomed the chance to build a general-purpose modeling system with both a classical and quantum-mechanical base. He notes, "That has come in handy, since my quantum-mechanical background has provided a level of rigor that might otherwise be lacking in some of my applications."

While Joe enjoyed his time with the army, he wasn't sure he wanted to make a career of it. There was a little too much risk of funding loss and significant mission shift. Through a recruiter, he moved to Marion Laboratories (a small search and development pharmaceutical company), where he performed similar tasks.

Joe saw a noticeable difference between working for the government and working in the private sector. He says, "The government was like a well-funded, somewhat mediocre university. We were tenured and could purchase hardware pretty well, but it wasn't always what I'd call state of the art. We did good work, and you could be more blue-sky since we weren't having to produce a profit—merely spend money wisely. There were a couple of negatives, though: The 'it's three o'clock; let's do it tomorrow' attitude wasn't something I was comfortable with. Also, PhDs with a little experience are near the top of the promotion chain, and I wasn't pleased with the prospect of only two more promotions in my technical career."

At Marion, Joe initiated efforts in computer-aided molecular design (CAMD), using both third-party and custom software tools; educated researchers in their use; developed software tools to support compound selection decisions; and helped investigate and rationalize molecular structure and reactivity.

Joe says, "I've been fortunate in that I've been able to change or expand my focus as I've followed the various interests and opportunities. I also think I've been able to be an integrator, trying to follow more modeling areas than some—I'm less interested in particular techniques, preferring to see how they're useful rather than how their theory can be advanced."

At Marion Labs, Joe assessed Ardent, Stellar, and SGI computers for purchase. Each had its own software to evaluate within a 20– to 40–day trial window. Testing their custom software, and porting his own code, meant that Joe worked closely with sales and support scientists at all three companies during the trials.

Ardent and Stellar merged at about the same time that Marion was acquired and the facility Joe worked at was closed, so he was able to move to Stardent and became their computational chemist. Joe not only developed and supported both chemistry and visualization software but also traveled to potential customers' sites to give product seminars and sales training.

Wavefunction was one of the quantum-mechanical software codes that Joe had ported to the Stardent system. He explains, "I made their code run, and then run faster, so I had sufficient credibility to get a job interview there when Stardent went out of business."

During his time at Wavefunction, Joe added a number of drug-design tools to their main product, Spartan. He continued to provide customer support and participate in tradeshows and conferences.

Over the next four years, the company gradually shifted its focus away from drug-discovery software and more toward quantum-mechanical and academic work, so Joe eventually decided to look elsewhere.

He knew several people from Tripos (a drug-discovery software company) from tradeshows and conferences, and several of them were familiar with Spartan and the work Joe had done there. When the right job became available, they thought of Joe. He accepted a position as a developer in the SYBYL software group and was eventually promoted to leading the group of 5–10 developers.

After spending some time as a manager, Joe realized he had been happier when he was doing the work than he was managing others who were doing it. He didn't want to go back to being a developer at Tripos, so he "returned to the lab" in 2001 and moved to the Abbott Bioresearch Center. He says, "I worked hard at Tripos to understand how one manages software projects, which has helped me husband the limited resources at Abbott. Once there, I decided to learn programming techniques in detail, and I have been fortunate to have the time and opportunity to do things I should have done 20 years ago."

Throughout all these changes, Joe tried to maintain a high profile in the industry and make sure people knew who he was and what he did. He explains, "I presented what I could at conferences, usually as posters, and tried to keep abreast of what was being developed where. Like all developers, I also tried to see what had people excited and tried to understand whether and how it could be implemented or packaged. Ours is a smallish field, so one's reputation must be protected. It really pays to make sure people know who you are, in a good way, and that you know who the important players are and what they are doing.

"Throughout my career, I've had to go to where the jobs were when they were open. We've lived on both coasts and in the middle, and we lack strong preferences. As long as housing's affordable and the commute's not too long, I'm satisfied.

"In pretty much all of my jobs, I've had a 'customer base' of one or more modelers to satisfy—PhD-level chemists who use the software I write. I've also been able to interact with 'real' chemists who have a computational leaning—physical organic or medicinal chemist types. I've usually interacted with or led scientific or computer-science developers, usually with tightly scheduled projects. I've worked with sales and support scientists, salesmen, and marketers, and spend a lot of time with sore feet in booths at American Chemical Society and other conferences."

On the whole, Joe feels he's been able to do what he's wanted to. He says, "I'm not sure whether everybody can say that, but that keeps things enjoyable. I think I've been well trained since leaving graduate school and have been fortunate to contribute to two well-received software packages. I'm far past programming for the fun of it—now the work has to be pertinent, something which makes it easier—or even possible—for modelers to help their projects, for it to interest me. In the olden times, computer fanatics 'won' if they created a code that was well known. In these terms, Kollman 'won' for Amber, and Zerner 'won' for Zindo. While I've not won at that level, a fair number of people have relied on my work in Spartan or SYBYL, and that makes me happy."

When making his career transitions, Joe was careful about how he crafted his resume. He explains, "Since I lack formal computer-science training—they were math courses back then—I've had to have a resume that had hard examples on it. For example, I've had people get demos of the commercial codes I've worked on."

He also believes it is important to know your own strengths and not try to know everything. He says, "I don't think it's possible to understand all there is about modeling and its application to medicinal chemistry. I've not had interest or time to understand organic chemistry, and what medicinal chemistry I've picked up has been in terms of tools and techniques. Those who know these areas usually aren't all that interested in software development, so it works out well."

Salaries seem to be roughly equivalent for development and laboratory work. He notes, "The salaries, naturally, are higher on the coasts and lower in the middle, which is a problem that people leaving school tend to not think about—the 'cost of living' varies widely. I think the software jobs pay a little less and I think biotech does as well. Both might sweeten the deal with non-salary compensation. Academia and government pay much less, but there are a whole lot of benefits once one is 'tenured' in either situation."

For the future, Joe feels that there are a limited number of positions that make sense for him. He could become a technical fellow, group leader for a development group, or head of development for a software company. "While I've done a little modeling, my strengths are in tool development rather than use, and that's where I want to stay," he adds.

"However, there are 10–100 times as many jobs as an applications modeler compared to as a developer. Thus, if one wants to do what I do, they've got to be good at it and, more importantly, they've got to be known and seen as good at it. Unless one is considered very good, they're going to find it hard to find positions. There's been a dramatic contraction in the number of commercial-software companies, and hardware companies can't afford specialists anymore. I think there's going to be a slight increase in development done at pharmaceutical and biotech companies but not enough to call it a great career option. I really appreciate the opportunities I've had to get experience in running a development group, and now really learning how to program well and at a higher level. But," he points out, "the bottom line is, any software application will have many more users than developers."

"That being said, somebody has to write the new programs and update the tools—they don't grow on trees. Computers have changed so dramatically, so existing tools are somewhat limited—and certainly inefficient. Now, we've got virtually infinite computer power, but are becoming limited by business details—licensing schemes, for example—or methodology limits. Anyone graduating with both programming and modeling experience, particularly from groups that have maintained large code bases, is quite desired by industry."

Joe adds, "Of late, I've tried to be a resource for younger folks I meet at conferences. They're starting out, and while their training's better than mine was, I think I can help them learn what *not* to do. Someone should learn from my mistakes!"

If Joe had to it over again, he says, "I would have worked harder at participating in national and even international groups than I have. Rather than merely going to conferences, I would have tried to do more organizing of them. I've thought 'I've got more important things to do' than schmooze, but I think that's been the wrong attitude. I'm trying to see how I can get back to maintaining a high profile in a correct manner."

Advice

Joe says, "I recommend learning how modeling tools are applied, as well as medicinal chemistry and even-further-from-modeling fields like immunology. Those with a more general background can look at bioinformatics or cheminformatics, but they're different areas with different needs. The former is well serviced by open-source software, so I'm not sure where the paid positions are outside of academic or government labs.

"If you really want be a software developer, learn to write—papers and grants are the coin of the realm, particularly in academia. Modelers are lucky in that they can publish with nonproprietary data and so can get far more papers out than 'real' chemists. I never did this well or diligently enough.

"Don't go into this field unless you're willing to work hard at it and run a few risks finding work. It's safer to be part of the customer base, as there's a whole lot more of them. The modelers, it seems, need to know far more than modeling. Of course, this seems to be the same situation facing organic chemists, since they don't seem to learn medicinal chemistry in school either.

"Learn to present oneself—at a poster, giving a talk, one-on-one, and so on. I've seen examples where people's reputation as a jerk from school or a postdoctoral position has haunted them for years. There are just too few of us and we intercommunicate far too often to not have people skills.

"Don't be too fond of particular technologies, languages, and so on. The bottom line is that modelers have to help projects succeed or succeed faster. If this isn't done, the programs and tools don't mean squat. I'm not sure people coming out of school have this understanding or perspective.

"On the technical side, understand how to run on multiple-computer or multiprocessing environments. Learn higher-level languages, since developer time is going to be the rate-limiting step in the future. Understand that code has to be stable and maintained—test suites aren't garbage; they're critical.

"Given the Web, I think young folks should create 'chemistry blogs,' particularly those discussing correct coding and development work. The only negative aspect of blogs is that it's way, way too easy to read and write garbage—it's all unedited. Blogs from 'somebody' tend to be good reading, but I'm not sure how to correctly transition from random person to 'somebody.'"

Predictions

As for the future, Joe says, "We're either facing a mature field where our customers have everything they need or an evolving field where we've got to handle complex computer environments and collapsing software pricing and revenue. A whole lot of work was done in the late 1980s on Vaxen that really should be redone. However, nobody's interested in paying for this.

"For example, if somebody determines how to score protein–ligand interactions, we're going to see a whole lot of money go to Dell, IBM, or Apple. . . . If each score takes eight hours to run, that's three a day. A cluster of 500–1,000 processors, which costs $500,000–1,000,000—or less, probably—will run 1,500–3,000 scores a day. This should be enough to help lead optimization. Similarly, if ADME and toxicity predictions work, iron's basically free. The software companies have to set their licensing charges and expectations correctly—nobody's much interested in paying $100,000 for software running on a $3,000 computer anymore."

He concludes, "Hopefully, this lull will end over the next few years. A number of us are trying to listen for anybody who gets oddly silent. We figure if somebody who should be talking clams up, they might be on to something. This will reveal itself via a productivity jump rather than publications—academia's not well positioned, as they have no data."

Vic Lewchenko

Assistant Project Manager (Contractor)
Pfizer, Inc.

BS, Chemistry and Math, Drexel University, 1976
PhD, Physical Chemistry, University of Wisconsin, Madison, 1981

Current Position

Vic Lewchenko is a long-term contractor with Pfizer, working as an assistant project manager. He is an employee of a staffing company but works onsite at Pfizer every day. His group of about 15 people is working to create and deploy a single global platform for data access and compound decision-making across research, enabling increased quality and productivity. This platform will replace a collection of legacy applications that were developed by companies Pfizer acquired over the years. As each company was acquired, its data and interfaces were added to the corporate system, resulting in a large number of disparate applications that do not always work well together. The goal is to provide a common framework and interface, so that all data and applications will work together, enabling scientists to access data more easily, learn more from that data, and make better decisions.

A large part of Vic's responsibility is to make sure the team is following the informatics software life-cycle process. This process defines deliverables and checkpoint reviews required at each stage of a software project's life cycle—including plan, design, build, qualification, and release. The deliverables are documents such as a requirements specification, project plan, test plan, communications plan, training plan, architecture scope, project schedule, and so on. Following this process ensures that all the many groups within Pfizer are aware of upcoming software releases and are prepared for the impact of a new version. These groups include the help desk, those that maintain the network infrastructure, business representatives, and many others.

Vic finds that working at Pfizer is "really like working in a commercial software company, except that the end users are employees of the same company rather than outside companies." He explains, "There is a software release schedule and software defects to be tracked, prioritized, and resolved. There are meetings with the various stakeholders. There is marketing and selling to the end users and to management."

Much of Vic's time is spent generating software-lifecycle documents. For this particular project, Vic prepared the requirements-specification document and project plan." He says, "These documents provide the details of what is being delivered—what functionality is being implemented—and how the project is going to be executed. The project plan also describes what resources from other Pfizer organizations will be needed, the project schedule, how deployment will take place, and so on." Thinking about the details of all aspects of the project in advance helps ensure that things will run more smoothly and provides advance warning for potential problem areas.

Vic follows up with developers to discuss the status of their projects and assigned defects regularly, to make sure they are still on track and on schedule. If they are not, he must decide whether to adjust the schedule, reallocate resources, or adjust the software specifications. Vic is responsible for investigating any reported defects in the system and assigning them to a developer for fixing.

He also meets with the various stakeholders regularly to keep them informed of the status of various parts of the project and to learn about any changes in priorities or requirements. When meeting with end uses, he solicits feedback on the software pieces that have been delivered so far, so they can be improved.

In the course of managing the project, Vic interacts with a wide variety of people, including quality-assurance specialists, software developers, chemists, biologists, project managers, software architects, technical writers, and many others. He says, "The ability to work as part of a team is very important for this type of work. Success depends on the coordinated efforts of a large group of people, a number of whom are offsite. The distance management makes it even more of a challenge."

Since the software Vic is building will be used to support drug discovery, it helps to know how chemists and biologists discover drugs, and the types of data they create and use. This helps him understand how the end users will use the product.

Career Path

As an undergraduate chemistry major, Vic was very interested in how computers are used to understand chemical systems and processes. He recalls, "One day in undergraduate organic-chemistry lab, I stuck a thermometer in my hand. This made me realize that lab work was probably not one of my strengths. Fortunately, our chemistry department had just hired a young theoretical chemist who introduced me to computational chemistry, and I learned there was a place for me in chemistry that would not endanger my health."

In graduate school, Vic's dissertation work centered on electronic-structure calculations. At that time, focusing research purely on computational work, with no lab work at all, was unusual.

Vic had two postdoctoral positions, both of which were the result of his responding to advertised openings. His first position after graduate school was at Johns Hopkins, developing a software application that combined quantum chemistry and model potentials to study small molecules of biological interest. This was a natural follow-up to his graduate work.

Vic's second postdoctoral position was at Washington University and was also purely computational work. However, here he looked at "a much different computational problem" from what he had done before. He explains, "For this work, we were applying statistical mechanics and treated proteins as amino acid chains, to investigate protein folding."

When Vic started looking for a position in industry, there were very few openings. Computational chemistry was a very young field at that time. He was fortunate that Tripos, one of the first companies to develop and market software for computer-aided drug design and molecular modeling, was located nearby and was looking for someone with his background.

Vic joined Tripos in 1984. He recalls, "Although I was hired as a scientific programmer, I quickly took on sales-support responsibilities. Within six months of starting at Tripos, I was transferred to the European sales office in Switzerland as the Tripos technical representative for Europe. I worked with the European sales agent to provide pre- and post-sales support and customer training. After 18 months I returned to the St. Louis office, and for the next several years I provided pre- and post-sales support for the United States, with occasional trips to Europe to help out the European office."

Vic eventually moved into product-management roles and then a release-management role for the company's flagship products. In 2005, he started as a contractor with Pfizer.

Vic has always been fascinated by the pharmaceutical industry and, in particular, the drug-discovery process, so he feels fortunate to be working for a leading company in this area, even though he is no longer "doing chemistry." He says, "The worst part is that I'm now managing the science, not doing it. However, I have been spending a lot of time looking at the data created by the chemists and biologists. All this data is chemistry-related, and it typically consists of a chemical structure, measured properties, and assay results."

For Vic, the greatest opportunity is "working with very talented people, both the software people and the end users, the chemists and biologists working to discover cures for diseases. The biggest challenge is keeping up with the rapid pace of change in technology, both on the hardware and software end and the chemistry and biology end."

Vic believes his future is in managing software development projects, he hopes in the life-sciences area where he can leverage his scientific training and where he truly enjoys working.

Advice

Vic cautions, "Keep up with technology and talk as many people as possible working in this field. I should have taken networking much more seriously earlier in my career.

"Having an understanding of both the information technology side and the chemical and biological side will give you an important advantage when seeking a job."

Predictions

For the future, Vic says, "I definitely think informatics is a growing area as companies try to tie together more and more sources of data and make this data available to researchers and others in the organization—patent, development, safety—in real time. Particularly as companies merge and acquire other companies, one of the big challenges is integrating the data from these other companies into a central informatics system. The creation of data and information will only increase and thus the need for creating systems to handle this will continue to grow. Also, the sources of the data are constantly evolving and changing—Laboratory Information Management Systems, screening, high throughput screening."

"There is room for chemists in this area who understand what is happening in the labs and in information technology. Someone needs to translate the needs of the chemists and biologists in the labs to requirements that can be implemented in software. Also, it is critical that the software support the workflow in the labs and not be viewed as a hindrance. Therefore, there is a need for someone who can speak credibly to both the lab scientist and software-development professionals."

John D. Clark

Associate Director
Accelrys, Inc.

BA, Chemistry, Frostburg State University, Maryland, 1980
PhD, Organic Chemistry, University of South Carolina, Columbia, South Carolina, 1986

Current Position

John Clark is an associate director, or supervisor, of software development in Accelrys's research and development department. Since Accelrys is a commercial, publicly traded software company, the work involves keeping software development on schedule, making decisions about how to implement specific new functionality, guiding software developers in how to make an easy-to-use software interface, answering questions from customers about how to use the software, and more.

John's typical tasks include determining requirements and predicting the level of effort needed to implement new functionalities, creating software-development schedules, then writing the software to implement new functionality and algorithms. He also designs software-user interfaces to meet chemists' needs for specific capabilities, and brainstorms with others to decide how to implement specific aspects of modeling molecules. For example, he explains, "We recently added the capability to model organometallic compounds that include dative and pi bonding." Once the new functionality is implemented, John helps test software before it is released.

John answers customer questions about how the software works and how best to use it to do a particular task. He trains internal scientists in the use of the company's products and trains software engineers in the software and on the chemistry aspects of their products. John is sometimes called on to write, edit, or review documentation for software his team develops.

Career Path

John got started in this career because "of a consuming fascination with computers, and modeling molecules on them." He explains, "For me it was a general interest in computers and building things—early on, I gravitated toward being a natural-products chemist because you build new things, and that is equally true in developing software."

John first saw a computer in the late 1970s, but he didn't connect it with chemistry. It was only during his graduate studies that he began to model different conformations of molecules and was able to develop software that could duplicate the conformations of crystal structures for some 14-membered ring natural products. From that start in graduate school, he continued to do informal computational modeling during his first postdoctoral position at Johns Hopkins.

It was while at Johns Hopkins that John decided to leave the laboratory and move fully into the field of computational modeling. His second postdoctoral position was at Washington University. He recalls, "There all of my work was done on the computer, modeling conformational transitions of helices. That work required a considerable amount of custom code, which I developed to complete the project—it was that work in which I discovered the enjoyment of making software. As a side benefit, this career change was certainly encouraged by my wife, who didn't particularly like me coming home from the laboratory smelling of ethyl acetate, tetrahydrofuran, and other organic solvents. As an experimentalist, I was as careful as anyone, but the exposure to those materials is virtually unavoidable."

When John decided to make a break from the typical path of bench chemistry, he sought out a variety of research projects to expand his skills in computational chemistry during his postdoctoral time. He was able to obtain a software-development position, at what was then called Biosym Technologies, through a recruiter. He recalls, "My original job title was 'scientific programmer,' and my job responsibilities were as a programmer, with expertise in chemistry. Even today, we hire a number of chemists into this kind of position."

For many years, John resisted moving into a more supervisory position. He says, "I didn't have any particular training in supervising other people and didn't really think that I wanted to do it. What I wanted to do was 'provide the vision for software products.' What I eventually realized was that the company didn't have such a position that wasn't supervisory, and perhaps rightly so. I don't really recall how long it took me to realize that, but, in retrospect, you have to be able to be persuasive and be able to lead people to be able to convince them to follow your 'vision' of what you want the product to be."

When the leader of John's group left the project for about six months, John became the de facto group leader. Shortly thereafter, his manager left the company, and John was asked to officially assume leadership of the group. He says, "From that point, I've been a supervisor of a development group, while at the same time staying involved with developing software myself."

In order to do his job well, John needs expert knowledge of both chemical structures and how molecules are modeled in software. He must know what constitutes the proper behavior in a computer model for small drug-like molecules and proteins, understand the implications and ramifications of the various approximations used, and also have a knowledge of software engineering and software languages. He points out, "Of course, understanding the intricacies of protein and small-molecule structure has been a key bit of knowledge, but more generally, I think the approach to solving difficult problems that is 'learned' in graduate school has been particularly useful."

As with any management position, John's primary responsibility is to keep his team busy and productive. He notes, "That requires planning ahead to a much greater extent than if you're doing the work yourself. The toughest part is when you have particularly productive people." His company has provided some training in management. But, he notes, "I don't find it particularly difficult to know what to do—managing people is not that hard if you can be reasonably empathetic."

The amount of time John spends actually developing software varies throughout the development cycle. Typically, he's involved with more planning and scheduling at the beginning, and as the release proceeds, he switches over to more hands-on development.

John's primary daily interactions are with the people he supervises in his group, three software engineers with mathematics degrees, and one PhD in theoretical chemistry. His secondary interactions are with upper management and the marketing group. The latter group determines the software's functionality, and they tend to be chemists as well. Less often, John interacts with customers, usually at pharmaceutical firms or in academics. He notes, "Having a degree in chemistry with lab experience is quite useful, since I can really understand the problems that chemists using our software have."

John feels that one of the greatest rewards in his line of work is knowing that the products they make are helping people everywhere. He says, "If I had become a medicinal chemist, I might have been involved with making a drug that would do that; however, virtually all drugs that are developed these days are developed using software at some stage. My involvement is more indirect; however, it's also more widespread. The challenge in doing this kind of work is that you have to be as credible in the language of software engineering and software-engineering principles as you are in the language of chemistry. One of the truly entertaining things is when I see an advertisement, or article in *C&E News*, or even in consumer magazines that use our software for the molecular illustration."

John feels there are other significant rewards in his career. He notes, "I can see that my efforts have directly affected pharmaceutical research, and the software that I've helped develop firsthand is used at nearly every chemical institution throughout the world. As an organic chemist, it would have been very unlikely that the results of my work would have had such a large impact. The worst thing about choosing this particular career is that I really enjoyed organic synthesis, and the problems that needed solving, and I don't have an opportunity to do that. However, I've taken up photography, so I can do such things as formulate new developers—an example of one of the few kinds of wet chemistry you can do at home."

Because working in scientific-software development requires additional skills relative to general organic chemistry, John believes the salaries are a little higher than for those who stay in the laboratory, depending on the current health of the software industry overall. From the standpoint of his company, his software skills make him equally valuable to a non-chemistry software company, and so the salary is commensurate with that paid to people in those positions.

Even though his position doesn't leave a lot of time for writing research papers, John has had the opportunity to get several patents for some of the software functionality he's implemented, which has been a valuable, though unexpected, experience.

If John could change his career history, he says, "I would have gone into the software field earlier. This is more of an option now than it was for me then, actually—when I was in college, there was really no such thing as a personal computer, and computer programming was really a specialized skill, so the choices are far more varied and plentiful now."

In the future, John would not expect to be involved in hands-on laboratory work again, having been out of the lab for so long. He suspects that in future positions he would primarily focus on software development in a scientific or medical field.

Advice

John advises, "In addition to excelling in your knowledge of chemistry, pursue knowledge in software engineering. As with many things, software engineering can be done at many levels of expertise, and for commercial-software applications, the importance of having a level of expertise in writing proper software can't be underestimated."

Predictions

"As the cost of making drug candidates, and doing actual laboratory work in any field, continues to increase, doing 'virtual' experiments will continue to have an increasing value," John predicts, adding that there will continue to be a need for new software tools and new functionality.

ADDITIONAL RESOURCES

American Chemical Society (www.chemistry.org) has an informative Web site; see especially the COMP division's pages

Computational Chemistry List (www.ccl.net), founded in January 1991, is a public e-mail list and Web resource dedicated to all things related to computational chemistry

EuroQSAR (www.euro-qsar2006.org) is a conference organized by the International QSAR and Modelling Society

Gordon Conferences (www.grc.uri.edu) is a prestigious international forum for the presentation and discussion of frontier research in the biological, chemical, and physical sciences, and their related technologies

Halford, Bethany. 2005. From Lab Bench to Laptop. *Chemical and Engineering News* 83 (16): 59–66.

Joel on Software (www.joelonsoftware.com) provides an introduction to software development

Lipkowitz, K. B. and Boyd, Dan B. 2002. Examination of the Employment Environment for Computational Chemistry. In *Reviews in Computational Chemistry 18*, ed. Kenny B. Lipkowitz and Donald B. Boyd, 293–319. New York: John Wiley and Sons.

MoBio (www.mobio.org) Missouri Biotechnology Association

Olin Business School (www.olin.wustl.edu) posts many events centered around the life sciences and entrepreneurship that are open to the public

RCGA Technology Gateway (www.technologygateway.org) has bioinformatics, IT, plant and life-sciences network groups that meet regularly

Slashdot (www.slashdot.org) discusses non-chemistry technical matters

Society for Information Management (www.simnet.org)

11 Chemistry and Education

SHARING YOUR LOVE of chemistry through teaching can make for a very satisfying career. Educators are needed at all levels, and an increasing number of people are making this a second career. Teaching is a career most people think is quite simple; in reality, education is a much more diverse career than most people realize, offering a satisfying and challenging profession with opportunities in the classroom, administration, and corporate training.

Teaching is very much a people-oriented career, with immediately tangible rewards. Instead of waiting years for research results, or for a product to get to market, a teacher knows in a few days—or even in one 54–minute class period, if he or she has managed to get the point across. At the end of a semester, improvement in students' understanding, and, one hopes, an appreciation for the subject matter, is apparent.

Teachers work closely with diverse student populations, in an environment that is highly structured from day to day but that allows flexibility year-round. Teachers know they are helping society in general, producing educated men and women who then go out and make their own contributions. In this way, a teacher's influence expands like ripples on a pond through the generations.

While many people think all that's needed to teach is knowledge of the subject matter, to be a professional teacher requires much more. Beyond a deep understanding of the material to be taught, teachers must also understand the learning characteristics of their particular student population (which depends on their age, motivation, background, etc.). In fact, many teachers feel that the most important characteristic in a potential teacher is love of the student groups they will be teaching. Other things can be learned, but unless the fundamental enjoyment of other people is present, the close, daily contact of teaching is unlikely to be rewarding.

Teaching can be an excellent second career. Professional certification programs exist to teach pedagogy (learning theory) for various age groups and subject areas, but there are some positions that do not require professional certification. These programs usually include supervised instruction—where the new teacher can learn by doing while being supervised by an experienced teacher. This ensures the best outcome for both sides—a solid grounding for the new teacher and a good experience for the students (who are not used as "guinea pigs" for an untrained teacher).

Although some level of educational background is required for any teaching position, adjunct positions at community colleges or primarily undergraduate institutions can be a good way to try out college teaching. The certification process required for high school teaching generally takes too long for it to be considered a short-term or interim solution.

* * *

THE most visible part of the education process is the teachers. Just about everyone knows what teachers do: they've seen them in action every day of their lives from the time they were four or five years old through early adulthood. But, most people have no concept of how incredibly satisfying teaching can be, how challenging it really is, or what a tremendous amount of behind-the-scenes work is required to make it look so easy. Many people are familiar with the expression "Those that can, do; those that can't, teach" and with the idea that teachers get low salaries and little respect. Many view teaching as something to do when everything else falls through or until something better comes along. Bob Becker, a high school chemistry teacher, recalls, "My own brother, who was graduating from medical school at about the time I was finishing my third year of teaching, asked me at one point, 'So, how long are you planning on doing this teaching thing, anyway?' I responded, 'I don't know, how long you planning on doing this doctor thing?' He didn't know what to say. He currently has a thriving practice—and I guess I do, too!"

Teaching takes place wherever learning takes place—in formal classrooms, small-group settings, one-on-one meetings, corporate seminars, high school classrooms . . . the list is endless. This means that those deciding to pursue a career in education can choose from a wide variety of places. Both high schools and colleges are frequently looking for teachers, and people with scientific training are needed in academic administration and corporate training as well.

There are many advantages to teaching as a second career. Having been out in the "real world" for a while, and especially having an advanced degree, generates respect from students. If they have to call the teacher "Dr. Smith," the teacher has instant credibility. (That is, unless proven otherwise by poor teaching methods.) An advanced degree may also help impress parents and can command a higher starting salary. Furthermore, work experience brings a depth and context to the classroom that brand-new teachers often do not have.

A teacher should always know more than the students, even at the end of the semester. Therefore, to teach high school chemistry, for example, a strong and detailed understanding of college-level, freshman chemistry is required. Depending on the school's curriculum, a basic knowledge of organic chemistry may also be required. Some chemistry courses are designed around specific applications, such as foods, industrial processes, nuclear chemistry, and so on, as a means to draw students into the course content. A curriculum like this would require that the teacher develop an understanding of the chemistry involved in each application. Furthermore, in many districts, a teacher is required to teach more than one subject, so not just chemistry, but also physics, math, or some other related subjects, must be studied and understood.

There are also topics beyond the subject matter that must be mastered by a truly excellent teacher. For an example, an in-depth understanding of various teaching strategies, the brain research related to learning, educational theories related to learning, current theories regarding assessment strategies and their implementation, the variety of student misconceptions about the subject area and how to change these, how to develop and implement lab experiences, and understanding how a students' background can influence their interest in or ability to learn the course content can all affect how a course is structured. In addition, new technologies including, podcasting, vodcasting, and chat rooms are changing the teacher's role in the classroom and opening up new ways for teachers and students to interact.

Student teaching is also a valuable part of preparation for a teaching career and is often required for an education degree. As a student teacher, one observes classes for several weeks then takes over the teaching, including lesson and assessment preparation, guided by an experienced teacher.

High School Science Teacher

High school teaching is both a career and a lifestyle, not to be entered into lightly. In order to be successful, you must enjoy the work and truly love hormone-laden teenagers. This career provides daily interactions with young people who are developing their own understanding of important scientific concepts and big ideas. Effective science teachers possess not only a thorough understanding of the subject matter but also a background in pedagogy and mastery of creative teaching methods that convey both the ideas and enthusiasm for the subject at hand. In fact, certification in these topics is required to teach in public high schools, and the certification process is usually too long to make teaching a viable interim solution.

Teaching physics or chemistry in a high school setting requires a significant amount of preparation time, in addition to class time. Demonstrations and laboratory exercises have to be designed, built, set up, and tested. Tests and quizzes must be prepared. Papers and lab reports must be graded. Virtually all teachers are expected to supervise some extracurricular activity, such as coaching, science fairs, clubs, and so on.

The best way to find out what a high school teacher does is to shadow a good high school science teacher for a while to see what they really do every day. Observe, analyze, practice, and reflect on what goes on, then decide if this is something that you might enjoy doing long-term. Most certification programs require some shadowing, and in the case of a full-time teacher-certification program, the final semester may be almost full-time shadowing a high school teacher and teaching part of their schedule.

Although private schools can legally hire non-certified teachers, more and more are requiring certification. Fortunately, there are a number of ways to become certified, some of which allow one to earn a living as a teacher while working toward the required certification. Requirements vary by state, and certification may not transfer from one to the next.

The monetary rewards for high school teaching are not as bad as most people think. Starting salaries are around $30,000 per nine-month year and can reach $70,000 per nine-month year in some districts. During the summer, teachers can teach extra classes for additional pay, find other employment, or take time off. Summers can also be used for professional-development activities, such as meetings and conferences. The National Science Foundation has professional-development Research Experience for Teachers (RET) grants to support K–12 and community-college science teachers' participation in research projects during the summer.

Just as salaries vary somewhat across the country, the demand for teachers also varies. Some states, such as California, Florida, Nevada, and Texas, are almost always looking for science teachers.

No matter the type of teaching job, the first few years are the hardest, according to most teachers. You are starting from scratch, and most new teachers are overly optimistic about what they can accomplish and how long it will take to do things. However, once they get through the first few difficult years, teaching brings deep satisfaction and joy to many educators, who can't imagine doing anything else.

Certification

There are many ways to get certified as a high school science teacher, and the appropriate method depends on the state in which you want to teach and your specific educational background and employment.

The "traditional" career path is to obtain an undergraduate degree in education, taking the chemistry and education courses you state requires. Typically, there are few required chemistry courses; they include an introductory chemistry course, quantitative analysis, organic chemistry and laboratory, physical chemistry and laboratory, and biochemistry. Often, only a few more courses are needed for a chemistry major. Other science courses may be required, such as those in earth science, biology, physics, and/or environmental science. Required education courses generally include those covering the nature of teaching, school organization, general teaching methods, teaching exceptional children, and methods of teaching physical science.

For candidates who already have a bachelor's degree in chemistry or physics, the required education courses can be squeezed into two semesters and one summer of full-time graduate work, if necessary.

Scientists who have a PhD in chemistry or physics can be hired by a public-school district to teach only that subject, with no additional training in education techniques. This route requires finding a district that is large enough to have a full-time chemistry teacher, for example, which is more likely at an independent, private school. There is little flexibility without additional certification.

Alternative Certification for High School Teachers

Many states now face a shortage of teachers, especially good, qualified science teachers. The need is expected to skyrocket with current teachers' impending retirements. In order to address this need, alternative routes have been set up to certify nontraditional teachers—individuals who have decided to change their career direction and become teachers after spending time in another profession. In many cases, having several years of real-world experience makes their teaching even more relevant for their students. Additionally, many corporate early retirees look at teaching as a way to keep intellectually active, contribute something to the community, and make a little extra money in their golden years. In fact, IBM has started a program to encourage those near retirement to move into teaching, by offering them partial salary and tuition benefits to do so.

Each state has its own requirements for teacher certification. Generally, the alternative routes require that the candidate have at least a bachelor's degree in chemistry, pass written and oral screening tests, participate in on-the-job teacher training (most often as a student teacher), and complete their pedagogy studies. The PRAXIS I PPST Skills Exam and PRAXIS II Subject Matter Exam in Chemistry are the most commonly used standardized tests.

In many states, if a district can document the unavailability of certified teachers in a particular subject area, they can hire someone who an undergraduate degree in that subject and a temporary or provisional teaching certificate. The person can then teach in that district while working to obtain certification. However, they remain at the lowest salary level until they complete the requirements for regular certification. By taking classes at night, on weekends, and during the summer, they can usually become certified to teach in two to three years while earning a living as a teacher. Often, those with a bachelor's degree can earn a master's degree while earning their teaching certification, a slightly more expensive option than taking only undergraduate education courses. In some cases, student-teaching assignments and internship requirements are waived or reduced for such candidates. However, unless there is a good mentor at the hiring institution, the new teacher will be learning by trial and error in the first few years, essentially using the first few classes as guinea pigs.

In some states, Transition to Teaching grants can be obtained from the Department of Education. These grants provide financial incentives in the form of stipends for mid-career professionals who want to get teaching certification. The grant is given to the state, which in turn makes funds available to candidates. Other sources of funding, such as the Hach Scientific Foundation scholarships for second-career chemistry teachers, are also becoming more readily available.

College Teaching

In addition to high schools, colleges are often in need of chemistry teachers. As opposed to large research institutions, where the emphasis is on research, at community colleges and primarily undergraduate institutions, teaching takes priority.

While full-time professors at most undergraduate-only institutions are expected to conduct some original research, it is usually scaled to the size of the grants that are available (much smaller for undergraduate schools than for large research-intensive universities), and much of the research time is spent training undergraduates to go out and conduct experiments on their own. Often, professors at undergraduate institutions design courses around their own research specialty.

Most college teaching positions require a PhD, but occasionally a master's level candidate will be hired for a part-time (and low-paying) position at a community college. Most colleges also prefer professors who have completed a postdoctoral position in which they conducted original research, so they can teach the students research methods from personal experience.

In addition to offering faculty positions, many colleges are increasingly hiring adjunct faculty. Part-time or adjunct faculty are brought in to teach a small number of courses (or even just one) for a semester, when there is not enough demand (or money) to hire a permanent staff member. Adjunct positions generally pop up quickly, when more students than expected register for a particular class for the upcoming semester and a new section must be added. There is often not enough time to advertise formally, so the positions are filled by word of mouth and often only a few weeks before the semester begins. An adjunct position will pay $2,000 to $8,000 per class per semester, depending on the type and location of the institution; they offer no fringe benefits. Adjunct professorships are on a semester-by-semester basis, and adjunct professors are rarely hired into full-time faculty positions. However, these positions are a great way to try out a college teaching career to see if it fits.

Another way to try out college teaching is to take a "leave-replacement position," filling in for a professor on sabbatical leave from a small college for a semester or two. This provides practical experience in the field, which not only looks good when applying for future positions but allows one to more thoroughly evaluate teaching as a career.

In additional to teaching, community college and undergraduate institution professors are usually expected to serve the department, institution, discipline of chemistry, and larger community. In many cases, service, along with teaching and perhaps research, are the basis of professional evaluations (and thus raises and promotions). The relative importance of each component will vary by institution, and understanding the ranking and adjusting your efforts accordingly can mean the difference between success and failure at a particular institution.

When evaluating opportunities, several things are important. First is the number of "contact hours"—how much time per week one must be in direct contact with students. The American Chemical Society's Committee on Professional Training recommends no more than 15 contact hours per week for a full-time position, but some schools require more. Another factor is whether or not teaching assistants are available to grade papers, which reduces the workload of the professor. However, using teaching assistants also means that the professor does not get clear, immediate feedback on what the students do and do not understand.

Profiles

Robert Becker

Chemistry Teacher
Kirkwood High School

BA, Biology, Yale University, 1983
MA, Education and Chemistry, Washington University, St. Louis, MO, 1990

Current Position

Bob Becker is a high school chemistry teacher at a suburban Midwestern high school that has 1,700 students and 14 science teachers. He teaches one or two sections of Advanced Placement Chemistry each year, along with three or four sections of Chemistry I. In his state, chemistry is required for graduation, and most students take it in their sophomore year. His class sizes vary, but most have around 23 students. The students he teaches vary widely in their level of motivational and competence. He has had "some pretty good students over the years." But, he notes, "I also have had quite a few who have no desire to be at school, let alone in a chemistry class."

Bob is an early riser. He prefers to grade papers and create tests before school, so he will get up as early as 3:30 a.m.. He has a 10–minute commute, which he shares with his daughter (currently a sophomore at the school at which Bob teaches). He refers to this commute as "excellent bonding time, before the sun comes up."

On Tuesdays, Thursdays, and Fridays he has a "zero-hour class" with his Advanced Placement Chemistry students. This is an extra session that meets from 6:55 A.M. to 7:40 A.M. and is required for students in AP science courses.

Before the school day begins, Bob usually sets up a demonstration or a lab. Then he starts a fairly typical teaching schedule. From 7:50 A.M. to 2:50 P.M. he teaches five classes, supervises one study hall, has one "free" period, and has 30 minutes off for lunch (with his fellow science teachers).

After the official school day, he normally has three to five students come by for individual attention or to make up a lab or test.

"This sounds as though it might get kind of dull and monotonous, but the truth is I *love* what I do. Joking around with the students, motivating them to do their best, getting to show off a huge repertoire of exciting labs and demos, mixing up how I present things, always developing new ideas and sharing them with others in the department. . . . It keeps the work exciting, and I never run out of new ideas to try," Bob says.

Career Path

Bob attended Yale University as an undergraduate, primarily because he was impressed that everyone was very involved in the community—much more involved than at any of the other colleges he visited. He majored in biology and was specifically interested in ethology (animal behavior) and psychology. He took just enough chemistry and physics courses to satisfy the requirements for his biology major and had planned to embark on a career in medicine.

Bob had done some creative writing while in high school, but it was not his strongest subject. In college, he took a few poetry seminars and was very proud of how his work was turning out. He sent some of his poems back to his high school English teacher, to show him how far he had come. Bob recalls, "He wrote back: 'This is great. If you are going to be on spring break next semester while we're still in session, how would you like to come in and present some of your work to my seniors—be a visiting poet, and teach them what you have learned?'" Bob thought, "Why not? So here I was, only two years older than these students, and I was their teacher—at least for the day. It went so well, about 10 minutes into it, I knew that it was what I wanted to do for a living!" Bob knew he wanted to go into teaching—math, science, maybe English, he just wasn't sure which subject. He went back to college and finished his biology degree while taking enough teaching classes to obtain his certification.

There were only two science majors in his entire graduating class who were considering education, and Bob believes that was a record high. He says, "Usually, it's zero per year. Most of my classmates were premed or pre-engineering."

As part of his education certification, Bob spent a semester as a student teacher, but, he notes, "Most of what I use I learned on the job. My student-teaching experience was pitiful: It was in biology, and it took place in an inner-city, all-minority school in New Haven; the school closed down shortly after I was there. When the principal learned that they were getting a student teacher, he immediately assigned my cooperating teacher to hall duty, so I only saw her once or twice the whole time I was there. Needless to say, it was not the best situation."

Upon graduation, Bob sent out letters to all the inner-city and suburban districts in the New Haven area, but no one was hiring. Eventually, he received offers from two districts—one as a full-time math–science substitute teacher in Darien, Connecticut (a very affluent, all-white district in the western part of the state), and the other a position in a one-room schoolhouse (K–12) in Tok Junction, Alaska. Bob took the road more traveled . . . he taught in Darien for one year and then moved to a neighboring district that had a considerably more diverse student population.

Bob was fortunate to have a mentor who helped him through the first few years. He remembers, "The first few years are the hardest for any teacher. You're starting from scratch: developing as you go. Plus, you really have no idea what works and what doesn't and how long to allow for various things. After the first few years, you can start coasting a bit. That allows you time to really enjoy it and spend more time getting to know the students, and other teachers, and getting involved in other aspects of the school."

His mentor encouraged Bob to take all the labs and demo ideas he was developing for his classes, write them up for publication, and present them at science-teacher conferences and workshops. Bob began doing that regularly, and it gradually snowballed into a second profession as a teacher of chemistry teachers. Bob has gotten hundreds of invitations through the years to present chemistry demonstrations all over the world. He has presented in at least 30 states, including during four separate visits to Hawaii. He has also made presentations in Canada and twice in Ireland. He very much enjoys doing these presentations and workshops, and they help make ends meet.

Bob taught in Connecticut for six years then decided to move back to Missouri, where he and his wife could be close to family and afford to buy a house. But even though he was certified to teach in Connecticut and had taught there successfully for six years, he was not even provisionally certifiable to teach in Missouri. He recalls, "As it turned out, Missouri didn't care that I was certified in Connecticut. I had to go back and take some completely ridiculous classes to get my Missouri certification. I decided to take a year off from teaching, take care of my newborn daughter, and finish my master's degree in chemistry while I was at it. The next year, I had no trouble getting work, since lots of schools were hiring." Bob has been teaching at his current school for 14 years and has no plans to change careers.

On a typical day, Bob interacts mostly with students but also with other teachers, counselors, secretaries, maintenance staff, substitute teachers, student teachers (interns), and parents. Typical tasks include lots of paperwork—writing and revising labs, worksheets, quizzes, tests, and note sheets, and sending them off to the copy center; filling out forms and progress reports; and always grading, grading, grading.

One of the most time-consuming forms Bob confronts is the individualized instructional plan (IEP). Students with learning disabilities, or those with special needs, each have an IEP, which describes in detail what accommodations the teacher is making to ensure the student is able to participate fully in the educational experience.

Bob also spends time mixing solutions and preparing equipment for demonstrations and labs. He is always looking for ways to improve his demonstrations and ensure they can be effectively used by the students. He is also constantly looking for new computer applications and often develops new PowerPoint presentations for class. He spends some time conferencing with parents and e-mailing other teachers and administrators regarding students about whom he is concerned.

After school, Bob supervises the Chemistry Club. The club began several years ago, when he started doing a tie-dye T-shirt lab with his students. Since it wasn't covered by the departmental budget, Bob had to charge the students, and he put leftover money into the science-activity account, from which he would then draw money to fund chemistry outreach programs. Eventually, he started the Chemistry Club to formalize these activities. They started selling tie-dyed shirts at the local Greentree Festival and donating the money raised to charity. They have also put on a spring demo show or two over the years, and one year took a weekend trip to the Museum of Science and Industry in Chicago.

This year Bob has assumed responsibly for Kirkwood Youth Service, a student service organization. Through this club, students work at soup kitchens, senior centers, blood drives, or elsewhere around town in volunteer capacities.

Currently, he is supervising a student teacher. He has done this about eight times, and many of his "disciples" teach in the area. He says, "Mentoring a new teacher is always a special challenge. To do it right takes a lot of time and energy, but the payback is enormous, not just for student teacher, but for the students, the profession as a whole, and the mentor as well. It also serves to remind me of how far I've come!"

Bob believes the best part of his job is the interactions with students. "They keep me young at heart, and knowing that every once in a while I make a difference in a student's life is the best feeling," he says. "The worst part is dealing with dishonest students, ones who cheat, or ones who just don't care, or ones who have terrible work ethics. Usually, these students are few and far between. Some years they are more prevalent, and I start to wonder if I chose the right profession."

Overall, for Bob "teaching just felt right." He says, "I can't imagine myself doing anything else."

Advice

Bob says, "In order to be a good teacher, you have to be a generalist and well versed in all aspects of chemistry. You must have confidence in your knowledge, confidence enough to admit when you really don't know the answer. Above and beyond all else, you must love children. Those who teach mold the future. If you truly love kids and can't wait to share your enthusiasm for chemistry with others, there is no better place to be than in front of a class.

"One thing about teaching that's unique: there are no promotions. That means that what I did my first day on the job, and what I will do my last day will be exactly the same—only infinitely better, I hope! Going into administrative work is *not* considered a promotion by anyone I know who takes teaching seriously. It requires quite a sacrifice to become an administrator, and the really good ones know they are really there not to be our bosses but to help us do our jobs better."

Predictions

Bob's biggest concern is that the "No Child Left Behind" legislation will lead to an overemphasis on testing and accountability. He fears, "We'll be so busy teaching toward the state-mandated tests that we'll have no time to teach our subjects. We'll be cranking out great test-takers, but no original thinkers. . . . Perhaps a bit pessimistic, but lately that is one of my main concerns."

As for national trends in the hiring of teachers, Bob says, "I know that every time we need to hire a science teacher, we really are competing with quite a few other districts trying to do the same. I'm glad that ours is a pretty desirable place, so we are usually able to hire top-notch people. I know that other districts aren't so lucky."

Donna G. Friedman

Professor and Chair of Chemistry
St. Louis Community College at Florissant Valley

BA, Chemistry, State University of New York, Buffalo, 1973
PhD, Chemistry, University of Missouri, St. Louis, 1979

Current Position

Donna Friedman teaches chemistry and chemical technology at St. Louis Community College at Florissant Valley. Her courses can include Fundamentals of Chemistry (chemistry for non-science majors), General Chemistry I and II (chemistry for science, engineering, and premed majors), Organic Chemistry, Analytical Chemistry, and Chemical Technology, depending on departmental needs each semester. In addition to her teaching responsibilities, Donna is also the chair of the chemistry department. Donna is unusual in that she has a PhD, whereas most people who teach at the community-college level have only a master's degree.

Although no two days are alike, a "normal" day for Donna involves both teaching and administrative responsibilities. Teaching duties can include teaching lecture or laboratory classes, meeting with students individually outside of class, grading exams and lab reports, advising students with respect to courses and career options, giving references and writing letters of recommendation for students, nominating students for awards, referring students to other departments or college services as appropriate, and learning to use new laboratory equipment.

The administrative work for the department chair includes activities such as supervising laboratory personnel; developing and submitting operating and capital-equipment budgets; approving expenditures and payroll; hiring part-time faculty; meeting with physical-facilities managers; monitoring waste storage and disposal; making sure safety policies are enforced in laboratory classes; meeting with students, faculty, staff, and administrators; reading chemical literature; evaluating textbooks for possible adoption; scheduling courses; and making teaching assignments.

Career Path

Donna was introduced to teaching in graduate school, when she was a teaching assistant. While she enjoyed teaching, she did not intend to make it her career. In fact, she was leaning toward applying for a position in an environmental laboratory.

After graduate school, Donna spent a year at Ohio State University as a postdoctoral research assistant, working on the synthesis of air-sensitive metal–carbonyl compounds of rhodium and iron. It was a one-year fellowship, following which she moved to St. Louis.

Upon arrival, Donna taught part-time at both the University of Missouri in St. Louis and at St. Louis Community College (SLCC) for a year before being hired full-time at SLCC. Donna recalls, "My decision to teach full-time at the community college was based largely on family issues. I had a young child and thought that teaching would offer some flexibility so I could spend time with my child." Teaching at the college level part-time allowed her to get her foot in the door, and Donna believes that she would not have been hired for a full-time teaching position if she had not had the part-time experience. Donna was hired as an assistant professor and was promoted to associate professor after three years, and then to professor after another three years. Donna has now been at SLCC for 25 years and has been chair of the chemistry department for the last 6 years. She plans to continue as both professor and chair for as long as possible.

Donna teaches not only chemistry but also chemical technology. The educational background of a chemical technician is less theoretical and more practical than is the formal training of a BS chemist, so the training requires a lot of hands-on time with modern equipment. Chemical technicians typically work in a laboratory, and most work in industrial settings, where they set up equipment, run reactions, and test for quality, performance, or composition of materials. Some technicians work on research projects, some go into sales, and some work in academic settings.

Donna always enjoyed teaching and interacting with people but also enjoyed working in a laboratory. Teaching chemistry and chemical technology allows her to do both. The best aspect of her job is that every semester is different—a different time schedule, different people to work with, different classes to teach. Each semester, she alternates between teaching chemistry-transfer courses and teaching chemical technology. The worst aspect of her career choice is that it is very time-consuming—courses are taught days, evenings, and weekends. If she's not teaching at night or weekends, she often takes work home. "Time off" between semesters is spent, in part, preparing for the next semester's classes and assisting students with registration.

Donna interacts with students, faculty, secretaries, administrators, student workers, and vendors. Her primary interaction is with students—male and female, ranging in age from their teens to their 60s, and having diverse cultural and educational backgrounds. Both native and foreign students attend the college now, so being able to communicate with all types of people is essential. Donna finds that "working with nontraditional students can be challenging." She points out, "The majority of students at a community college will work 20 to 40 hours a week and have family responsibilities. Many are 'returning' adult students. One must be understanding of the outstanding demands on students while maintaining high academic standards."

Community colleges in general have the same promotion track as research institutions for full-time faculty—instructor, assistant professor, associate professor, and professor. Professional development is important for faculty members at all levels. Faculty are expected to stay current in their field by taking courses; attending seminars, workshops, and conferences; and taking active roles in professional organizations, such as the American Chemical Society.

There are other opportunities for community-college teachers to expand their professional horizons. Teachers can also apply for administrative positions or work with faculty in other fields to develop interdisciplinary courses.

While salaries for those in this field may be lower than for those who remain in a traditional laboratory position, there are advantages. Those teaching above the normal course load in the spring and fall can get additional pay. Fringe benefits generally include (but are not limited to) medical and dental insurance coverage, retirement plans, and tuition waivers for employees and their immediate family. Summer teaching is often optional, leaving three months to pursue other interests. Many professors work in industry during the summer or take courses and attend seminars and workshops for professional development.

Advice

The skills Donna finds most valuable are her knowledge of chemistry and mathematics, writing, computer, and organizational skills, and the ability to interact effectively with diverse populations of students. She feels that performing undergraduate and graduate research, working as a teaching assistant in graduate school, co-authoring papers, and writing her dissertation were all valuable experiences that helped mold her into a good teacher.

Donna's advice to those thinking about a career as a community-college teacher is "try teaching part-time before applying for a full-time teaching position." She notes, "This will provide real experience that will come in handy when applying for full-time positions and will also give a real feeling for what the job is like—and let you decide if you want to do this full-time."

Predictions

Donna says, "There will always be a need for chemistry instructors. However, there are only a few chemical-technology programs in the United States, primarily because the laboratory instrumentation required to run the programs is very expensive. The programs must keep up with changes in technology, so capital and operating expenses are high. Therefore, there are limited opportunities to teach in this specific area." Donna does not foresee an increase in the demand for instructors for chemical-technology programs, even with an increased demand for chemical technicians, because most colleges cannot afford to start and/or maintain these programs. Class sizes must be kept low (16 is the absolute maximum at Donna's school) because a lot of individual hands-on laboratory instruction is necessary.

Donna also feels that "as student populations become more diverse, a teacher's communication skills become ever more important." She adds, "Student diversity affects class scheduling as well. Students often have their own schedules—work and family—that they have to schedule classes around, so the college tries to be flexible by offering classes on nights and weekends to accommodate students' needs. This may require some full-time faculty members to teach nights and weekends, or adjunct faculty may be brought in to teach those classes."

John C. Mackin III

Science Department Chair and Chemistry Teacher
Kirkwood R–7 School District

BS, Chemistry, St. Louis University, 1980
MS, Chemistry, St. Louis University, 1985
Teacher Certification, Chemistry, Secondary Level, 1996

Current Position

John Mackin is both the science department chair and a teacher in that department at a suburban high school. As such, his professional responsibilities are split into several realms: teaching students, mentoring new teachers, and managing departmental budgets, scheduling, and supplies. Working in a high school, his day is fairly structured. He is required to be at work by 7:30 a.m. and teaches four chemistry classes a day. One period each day is devoted to a "sixth responsibility," which is a duty that each teacher performs to help the school overall. This could include monitoring a study block or assisting in data collection and analysis.

As department chair, he has one extra non-teaching period each day to devote to administrative duties, in addition to his non-teaching teacher-planning period. (Teachers who are not department chairs teach five classes, with one sixth responsibility and one planning period.) He has a 30-minute lunch period, and as department chair he is responsible for planning and facilitating the weekly department meeting that occupies one lunch period each week.

After the school day ends, he tutors students who need help in chemistry, then grades papers and sets up labs and demonstrations for the next day. Typically, he stays at school until about 5 P.M. In his first few years as a teacher, he would grade papers and plan future lessons for about two hours each evening and for about four hours over the weekend. Now that he is in his eighth year of teaching, many of his lessons are well refined, and significantly less planning time is needed.

As a teacher, John feels that his primary responsibility is to develop a relationship with each of his students to help them maximize their learning potential. In the beginning of the year, he strives to learn students' names and some aspects of their personality that will allow him to start making a connection. As the year continues, he interacts with each student as he checks their work and assists them in their learning both in and out of the classroom.

Career Path

John started his career with a BS in chemistry then got an industrial product-development job. He earned an MS at night, which took him five years. He subsequently moved to another company for a job with more product-development growth potential. After 11 years with the second company, John realized he was ready for a major change.

He was traveling often and was bothered when his children began asking, "Are you leaving again?" John also noticed problems in his company's future. The firm was attempting to significantly broaden its product line with inadequate funding and with insufficient dedicated production personnel, which made his position highly stressful. (In fact, within a few months after he left the company, there was a reorganization to correct those deficiencies, which may have eliminated his position.) Furthermore, he wanted to contribute more to society than he could at that job.

In thinking about what he wanted to do, John realized that his industrial career had given him some opportunities he enjoyed. He recalls, "I had been involved in assisting in the growth of subordinates on numerous occasions, and realized that I enjoyed this aspect of my job. During work on my master's degree, I helped other students with the coursework and found this gratifying as well." John decided to pursue teaching as his next career. He estimates that only about 25% of current teachers returned to the classroom after an industrial career: most teachers choose teaching as a first career.

To enter the teaching field, John had to earn a teaching certificate. He chose to teach chemistry, since he already had a master's degree in chemistry and could most quickly complete his required certifications using this degree. The certification required two semesters of evening classes and two full-time semesters, one of which was student teaching.

John obtained his current position when one of the chemistry teachers at the school at which he was student teaching chose to leave. In 2003, the principal asked him to become the department chair based on both his teaching and management abilities.

To assure a consistent learning experience for all students of chemistry regardless of their teacher, John collaborates with other teachers at workshops several times a year to develop curriculum objectives. They use state and national standards, guidelines from various national science organizations, and local input. These objectives are updated annually as state and national standards are refined and feedback from the previous year is incorporated.

Using this curriculum as a planning guide, John develops daily lessons for his classes along the continuum from teacher-centered (guided lectures and modeling) to student-centered (labs, projects, group inquiry) activities. Lessons are planned to incorporate the latest information including learning from brain research, the influence of multiple-intelligence theory, attention to student learning styles, and content-specific teaching strategies. The lessons not only center on content but also on specific process skills such as experimental design.

For example, one lesson involves having the students complete a cookbook-type lab on flame tests using Bunsen burners and various metallic-compound powders. Afterward, they discuss the experiment and design additional experiments of their own based on questions the initial experience prompted. This design experience gets them thinking about variables, procedures, data collection, and so on, and they have fun.

Preparing for these lessons involves developing materials such as guided-inquiry sheets, strategy-modeling approaches, skill-building problem sets, follow-up questions for laboratory activities, project development, demonstration materials, and inquiry or experimental design labs. The option to include technology must also be planned and can involve PowerPoint presentations using a laptop computer and projector; laptop computers for informational research, data analysis, and/or lab simulations; and a flexible camera for zooming in on small-scale materials.

Along with the preparing for class, John must develop assessment rubrics and other tools to ensure that students have learned the material. These tools include both quizzes and cumulative tests. The assessment questions range from multiple-choice to short-answer to performance, often including a lab component.

As department chair, John is responsible for guiding new teachers entering the department and for overseeing budgets, scheduling classes, and procuring supplies. This amount of time required varies throughout the year. In the fall, he must observe new and inexperienced teachers and assist them as needed to help them develop their teaching skills. In January and February, he prepares the budgets for supplies and textbooks. Throughout the year, he monitors spending to assure it remains within the budget. In February, he develops the teachers' schedules for the next school year, based on the number of students enrolled in the various science classes. Multiple permutations of possible schedules are developed for the science teachers' review and approval. John also attends weekly meetings of all department chairs and administrative personnel. The weekly science-department meeting allows him to disseminate information from the chairs' meeting and allow for discussion and feedback on the issues.

John feels that his industrial and management experience has allowed him to more easily see the big picture, education-wise. This has allowed him to contribute to the educational process beyond the classroom, through interaction with administrative personnel. He has been involved in assessment committees, vision committees, and data analysis, all of which help administrators guide and direct students and programs.

John says, "My 'real world' experience gives me a perspective on my content that a teacher who has never worked outside the classroom does not have. I can more effectively talk to students about the uses of chemistry and how it impacts their lives." He observes that his work experience gives credibility with the students, and notes, "My industrial management experience also positively impacts how I plan my teaching, organize my information, and see the bigger picture to the benefit of my department and myself."

John most enjoys the opportunity to continue to learn about teaching and chemistry through researching, developing, and executing lesson plans. He also derives great satisfaction from seeing students recognize their potential and become better learners themselves.

For him, the worst aspect of teaching is the frenetic schedule. Working in 54-minute blocks while repeating lessons, doing the job of department chair, and other undertaking other duties is stressful. Another downside was the reduction in salary—"teaching pays about half" what a corporate position pays, "with fewer benefits," John notes.

As for his future, there are few career-growth opportunities for teachers. Beyond classroom teaching, there are administrative positions, including subject-facilitator, assistant-principal, principal, and district-superintendent positions. However, most of these jobs require that one have additional education and state certifications and be willing to forgo direct interaction with students.

Advice

For those considering a transition to teacher, John suggests, "Supplement your education degree with real-world experience in your field. This will make your teaching more authentic."

He adds, "The greatest opportunity of teaching is leaving a legacy of changed lives. Every time I interact with a student, I can positively—or negatively—influence their future. This may not be apparent today, and I may never know what impact I have had, but I believe I make a difference. However, the challenge is educating *all* students in whatever subject matter you are teaching. Some students are difficult to reach, either due to lack of skills or lack of interest. Trying to reach these students and not succeeding is the most frustrating part of teaching."

Predictions

Looking toward the future, John says, "If the federal government continues to expect that *all* students will learn—including students with severe disabilities, and the evaluation of this learning is through inadequately designed, state-mandated assessments, teaching will be more and more constrained as teachers attempt to educate students in ways that will reflect positively through the assessments. This will lessen the joy of teaching and cause current teachers to reconsider their career choice.

"Additionally, as states increase the educational requirements for teachers as a way to potentially assure greater student learning, obtaining a teaching degree will require five years of college, and fewer prospective teachers will choose teaching as a profession."

Arlene A. Garrison

Assistant Vice President for Research
University of Tennessee

BS, Electrical Engineering, University of Tennessee, Knoxville, 1988
PhD, Analytical Chemistry, University of Tennessee, Knoxville, 1981
BA, Liberal Arts College Scholars, University of Tennessee, 1975

Current Position

Education takes on an added dimension at a major research university. The research laboratory is an additional venue for student enrichment. Students at the undergraduate, graduate, and postdoctoral levels participate in original research projects, where the outcome of the research is uncertain. While there are parallels to classroom and teaching laboratory-learning experiences, the research laboratory requires many new skills such as in depth critical thinking and analysis and greater emphasis on practical skills.

A university with a strong focus on research, in addition to the traditional teaching and public-service missions, requires additional skills from instructors, and many more administrators. Arlene Garrison is the assistant vice president for research at one such institution, the University of Tennessee. Her main responsibility is working with faculty to prepare and submit proposals for external funding for research projects. The University of Tennessee receives about $250 million per year from external sources. Arlene points out, "Demonstrating high ethical principals as we comply with all regulations is essential to our externally funded activities," and this is one of her major concerns.

Much of Arlene's time is spent in meetings. Teams preparing proposals usually meet several times to write and review the documents. Administrators meet to learn about new regulations and see that the processes are implemented uniformly across the campus. Sponsors sometimes visit the campus, but normally meetings are held at the sponsors' headquarters. Arlene travels to the state capital (Nashville) and to Washington, DC, several times a year. When she is not in meetings, she is generally responding to e-mail, generating e-mail, or talking on the phone. Like most administrators, Arlene finds it necessary to work late or on weekends to find the time to plan for the future.

Interaction with her professional colleagues is also a significant part of her normal work routine and includes professional meetings, proposal reviews, and committee work for professional organizations. Arlene believes that "professional societies are important as a public voice for science." The American Chemical Society is the organization with which she is primarily involved, and she is active in both the local section and on a national committee. The contacts she makes there often help her find solutions to her work challenges, and the friendships formed there are an important part of a satisfying professional career.

Arlene has some routine responsibilities as well. All research contracts have to be signed by one of three people, and the responsibility is coordinated with meeting schedules. Reports have to be generated, including research outcome reports and financial reports.

Selecting the winners of various types of internal research competitions is a more pleasant responsibility. Students compete for money to fund a project or to match a federal grant. There are many situations in which resources are limited and requests are many. All selection processes include representatives of all the stakeholders, and the process can become quite complex because fairness is a priority.

In recent years, the work climate has changed. All organizations must be flexible and dedicated to continuous improvement, and the University of Tennessee is no exception. A surprisingly large amount of Arlene's time is spent studying benchmarks, discussing the future, planning better processes and policies, and implementing changes.

She says, "When I come to work, I rarely know what will be the focus of my day. A large part of my job is fixing problems. This includes handling contract-negotiation issues, finding answers to questions, and finding the right folks to get answers."

Arlene talks with other research administrators constantly, and often communicates with faculty and research sponsors. Most of her regular interaction is with highly educated professionals with varied technical backgrounds who travel widely. The diversity of these interactions has increased over the years and will probably continue to increase in future years.

Career Path

Arlene recalls, "I chose chemistry because of my interest in the components in a mixture. In middle school, I learned that chemists held the keys to solving the mysteries of materials. Why did water in different cities taste different? What was in paint that smelled bad as it dried? Why did adding soda and vinegar to cake batter cause bubbles? In the ninth grade, I did a science project that used acid to break down wood into sugar. The science was developed in World War II to provide a replacement for sugar supplies. Later, I learned that many major breakthroughs in chemistry have happened as a result of shortages of natural materials and that the reason for the shortage was often war and the related embargoes. In college, I found myself fascinated more by the measuring of the components of a mixture than by the creation of new materials. Sensitive instruments are available to measure small amounts of unwanted substances in a very complex mixture, and the analysis requires a lot of expertise."

Despite having her senior year in college interrupted by a car accident and lengthy recovery, Arlene went on to graduate school. She earned a PhD in analytical chemistry and began a postdoctoral position in a research laboratory on the path to a fairly traditional career. She recalls, "Family and health issues impact all careers, and mine was no exception. My initial job after having my son was selected to allow some flexibility in scheduling."

After some time in research, Arlene developed an interest in research management. Research administration is a complex process because of the many research regulations. During the course of a big project, Arlene's project leader left the university, and she was put in charge temporarily. The position was well above her former position, and Arlene had to quickly learn many new skills.

She recalls, "I suddenly managed an office staff and was responsible for financial matters such as budgeting and tracking expenses. I also had primary responsibility for recruiting corporate sponsors for the research program. Contract negotiation was probably one of the most difficult aspects of the new position. My former supervisor was very helpful in providing a quick review of most of the processes, and he continued to be available by phone for several months. Universities assume that any intelligent person can manage, and they have no formal program for new administrators. A couple of the industrial members of the center provided significant help in my transition by making portions of their internal management-training programs available to me. The hardest thing to learn for a new administrator is delegation of responsibility. The mixture of on-the-job training along with a higher level understanding from the formal training worked to make a quick transition from researcher to administrator."

Arlene became interested in the planning and reporting side of major research projects. She also became very frustrated with the rules of the game—the institutional policies and procedures. At that point, Arlene decided to shift her career direction.

She says, "I chose to pursue a career as an administrator in order to influence the allocation of resources in pursuit of higher goals for society and the university. It is typical for individuals who have been successful to advance to administration. Reasons for the transition vary and can include higher salaries, the opportunity to have a long-term impact, and a perception that the work is less time-consuming. Laboratory work can be very demanding and often must be done at a particular time in a particular location. It is not a career that adapts easily to flex time or working from home. To the researcher, it can appear that administrators have a light schedule and just sit in their offices. In reality, administrators work very long hours with a lot of travel to visit sponsors and also have meetings and send e-mail during evening hours."

Later in Arlene's career, the University of Tennessee underwent a massive change in leadership, with five presidents in five years. There was a corresponding change at the vice-presidential level almost as often. The rapid change provided many unexpected opportunities to explore new aspects of administration and to work with new colleagues.

Arlene recalls, "With the rapid succession of new presidents and vice presidents, I was introduced to the world of process analysis and organizational redesign. Each new boss had experience with a different way of doing things and believed that a change in the organization would improve productivity, so change was implemented. Corporate organizations do this often and usually employ consultants to analyze and manage the process changes. Universities tend to hire a consultant to help with the analysis and then use internal folks to design and manage the changes. The process of change requires making a detailed map of duties and people in the old organization, which are then analyzed to identify redundancy and ways to improve efficiency are identified. Some activities are centralized and other duties are outsourced. Since some duties are intentionally dropped, staff reallocations are required. A map relating duties in the old and new organizations must be developed. The biggest challenge is watching for any duties that may be accidentally dropped.

"The activities related to change management were very interesting to me. I learned it was critical to empower and motivate the researchers and administrators. A well-designed process, with a focus on the desired outcomes, can lead to a positive change, where everyone feels their contribution is important. In the many reorganizations, I was tasked with work related to compliance with regulations. Regulations dealing with treatment of human subjects, animals, and other critical materials are very important in research administration. Ethical university researchers are concerned with the real issues in compliance and are committed to appropriate behavior. The necessary paperwork is very time-consuming, and maintenance of appropriate paperwork is very difficult because of the many and varied regulations. It was very interesting to learn about the various experiments and the degree to which researchers were concerned with care for animals and similar issues, but the portion of this role that required enforcing paperwork was not very satisfying."

In her current position, Arlene helps create policies that support the researchers, and she can help inform faculty of the reasons for the rules. With better policies and a better appreciation of the policies, research at the university works more smoothly.

Arlene notes, "Skills I learned in chemistry classes are essential to the job I do today. In large part, the most valuable skills are mental and organizational rather than technical, but there are parallels. Chemical reactions occur in steps, and each step can be optimized, much like organizational processes. In order to avoid safety risks and save time, chemists review prior work in a field before starting work. Similarly, administrators routinely benchmark processes developed at other schools and read regulations very carefully before developing new policies. Critical thinking is probably the most important skill. Understanding how to break down a large project into steps and how to schedule those steps is also very important."

In any administrative role, the hardest daily challenge is clear, unambiguous communication to everyone involved. The most significant challenge is assuring compliance with regulations, especially rules involving finances or the use of human subjects and animals.

The most interesting part of her career has been the chance to travel and spend time with a wide range of people from all over the world. She recalls, "My first lengthy business trip was to a NATO-funded institute in Italy for two weeks. At that meeting, the top scientists in my field spent time together in lectures and discussion starting early in the morning and ending in less-formal conversation after dinner. After more than 20 years, I still count many of the attendees as friends. I have had many similar opportunities and have founds ways for my family to travel with me to many interesting locations."

As for compensation, in a simple comparison, administrators receive measurably higher salaries than bench chemists, roughly 15–40% higher, depending on circumstances. However, when all sources of income are considered, faculty often receive a salary comparable to that of administrators because faculty members have a number of potential sources of income not available to administrators, including summer employment and consulting.

Research administrators can advance to higher level university positions such as provost or president. They can also move to similar positions in government or at private laboratories. Some research administrators have found outstanding opportunities working with high-tech startup companies. However, progressing in the field of research administration usually requires long-distance moves.

Advice

Arlene notes, "Life involves many changes. Any change in circumstances can be approached as a challenge or an opportunity. At some points in my career, I made excuses for changes in direction, even though the decision was mine. My recommendation to those at an earlier career stage is to accept decisions and take full responsibility for them. The immediate outcome is the same. The positive attitude enhances the experience going forward."

"If you're thinking about administration as a career, first establish a good research career in order to gain experience and respect from colleagues."

Predictions

"The field will experience growth for many years in the future," Arlene observes, adding, "Regulations and ethical concerns are only expanding and administrators are necessary to manage these issues. More universities are interested in performing research, and there is greatly increased interest in the potential for economic development, such as new company starts based on university-developed science and technology."

ADDITIONAL RESOURCES

Association for Assessment and Accreditation of Laboratory Animal Care International (www.aaalac.org) promotes the humane treatment of animals in science through accreditation and assessment programs

Association of University Technology Managers (www.autm.net/index.cfm) deals with economic development and university spin-off companies, an important and expanding part of research administration

ChemEd-L (mailer.uwf.edu/listserv/wa.exe?SUBED1=chemed-1&A=1) is a chemistry education discussion list

Chem 13 News (www.science.uwaterloo.ca/about/newsletter.html) is a magazine for teachers of introductory chemistry courses

Council on Research Policy and Graduate Education of the National Association of State Universities and Land-Grant Colleges (www.nasulgc.org/councils_research.htm)

Chronicle of Higher Education (www.chronicle.com) posts employment listings

Hal's Picks (www.umsl.edu/~chemist/books/halspicks/halspicks.html) lists books on topics likely to come up in introductory chemistry courses

Journal of Chemical Education (jchemed.chem.wisc.edu) is the premiere journal for chemical education research

National Center for Alternative Certification (www.teach-now.org; (866)778–2784) provides information on alternative routes to certification in all the states

National Center for Education Information (www.ncei.com) is a national source of information on education and teacher preparation

Reed, John H. 2006. *A Guide to Classroom Instruction for Adjunct Faculty*. 2nd ed. Washington, DC: American Chemical Society. For purchase at www.chemistry.org or by calling (800) 227–5558.

Science (www.sciencemag.org) posts scientific articles of general interest

Schwartz, Truman; Archer, Ronald; El-Ashmawy, Amina; Lavallee, David; McGuire, Saundra; Richmond, Geraldine; and Eikey, Rebecca. 2006. *And Gladly Teach: A Resource Book for Chemists Considering Academic Careers*. Washington, DC: American Chemical Society. The booklet, available at www.chemistry .org, covers the decision to pursue an academic career, preparation for an academic career, the search for an academic position, and keeping an academic appointment.

Society of Research Administrators (www.srainternational.org/sra03/index .cfm)

Transition to Teaching Grants (ed.gov/programs/transition.html)

Wisconsin Standards for Teacher Development and Licensure (www.dpi.state.wi.us/dpi/dlsis/tel/stand10.html)

12 Chemistry and Everything Else

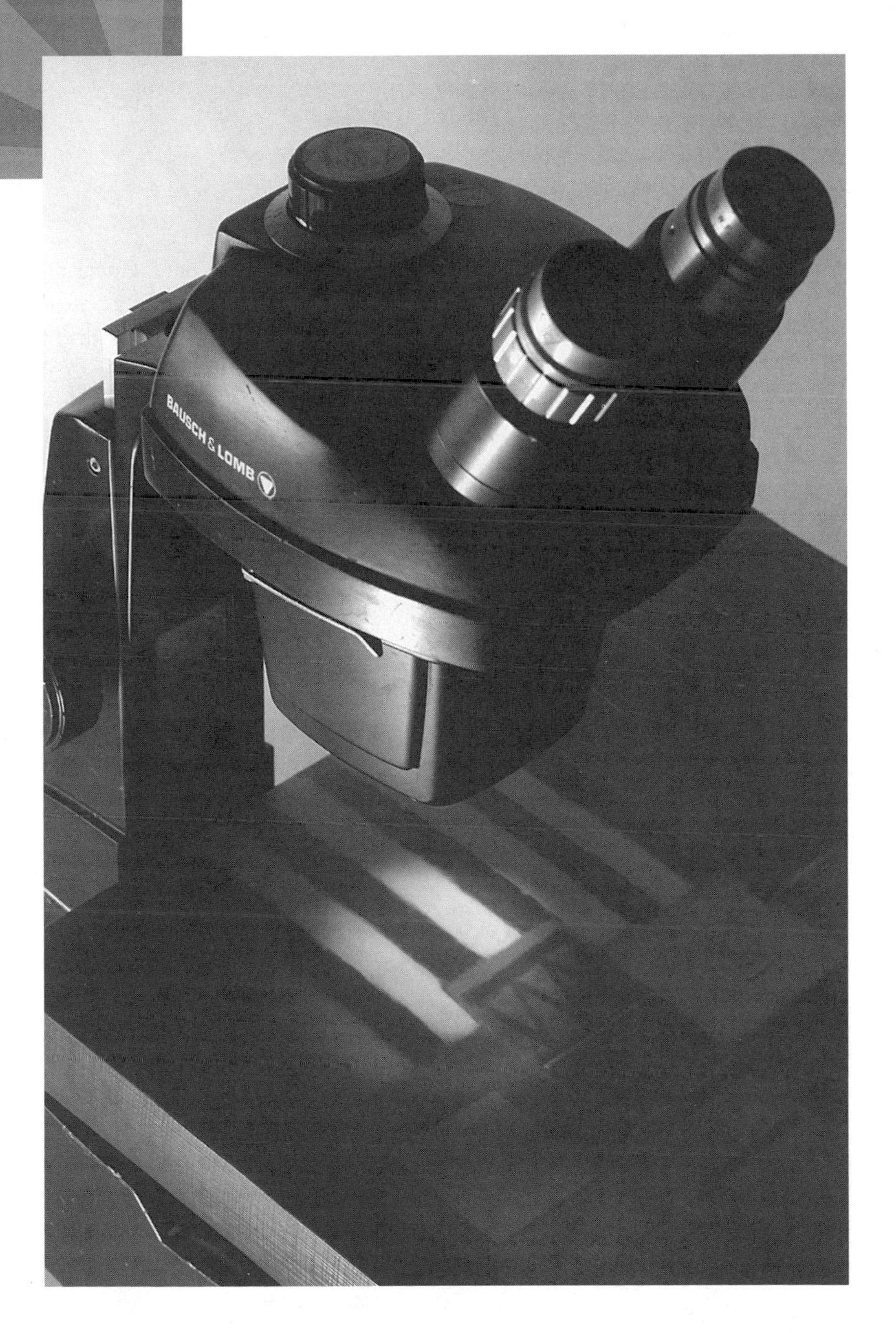

CHEMISTRY IS EVERYWHERE—in materials and their pigments, in how things are created and how they are used, in how they fit together and how they affect the world around them. By looking around, all sorts of interesting niche markets can be found—or created!—for the application of chemical expertise. While these careers may not be for everyone, they will certainly provide food for thought on what is possible, and, ones hopes, spark even more ideas.

Chemistry and Imaging

Graphs, spectra, and charts are often used to convey scientific data. However, scientific concepts are sometimes better communicated through carefully crafted images. Detailed, accurate images not only convey huge amounts of data in an easily understandable format but can be used in marketing to give technical products more credibility with scientists. More than creating mere "eye candy," a photographer or graphic artist who understands the science behind the product can create an accurate, compelling image that will engage scientists, not offend them. These images not only aid understanding of complex concepts but also bring the beauty of science to the general public. "Good scientific images can create awareness of the beauty and wonder that exist in our everyday lives and educate people about the knowledge that science provides," says Dana Lipp, a science photographer.

Many scientists document their work with photography but do not have any formal education in that field. With the advent of digital imaging and computer manipulation, there are more ways than ever to crop, merge, color, and otherwise edit images. Color can be added to highlight and enhance specific aspects, and multiple images can be combined and cropped in any number of ways. Presentations and publications would be improved if the images used were not only scientifically accurate but also well composed and visually stunning. However, this task brings with it the challenge of making sure the edited image still represents the data accurately and that modifications to parts are clearly described. Visual appeal and scientific accuracy must be carefully balanced.

Both chemistry and photography are about seeing patterns and connections between things and finding out why they are there. In science photography, the creativity comes in looking at things in a new way.

Science photography is a very small niche market, and most photographers who have full-time employment are in academia. For example, a small number of universities hire photography experts to collaborate with scientists and engineers to create images for journal submissions, publications, and presentations to other scientists and the general public.

Science photography is just starting to gain attention from the general public. The Visions of Science Photographic Awards competition is held annually in the United Kingdom. Entrants compete in a number of categories; the winning images receive a total of £7,000 in prize money and participate in a touring exhibition that travels the United Kingdom the next year. The fifth annual competition, held in 2005, garnered about 1,300 entries. Unfortunately, the contest is closed to foreign entries.

Chemistry and Materials

One of a chemist's traditional jobs is creating new materials. But once those new materials are created, they must be turned into a useful product if they are to produce a profit. The product may be a drug, a better cleaning agent, or in some cases, a new fabric with specific electrical properties. Many different specialties are involved in transforming a laboratory curiosity into a marketed product.

Chemistry and Art

Most chemistry laboratories have at least a little art—since 1967, the cover of every edition of the *Aldrich Handbook of Fine Chemicals* has shown a different masterpiece with every edition. In some laboratories, art is the focus of the chemistry. The most common combination of chemistry and art is in the identification and preservation of historical artifacts. Conservation of paintings and other works of art requires a three-way collaboration between conservation scientists, conservators (who do the hands-on treatment), and art historians or collection curators.

Conservation scientists help conservators understand the materials and methods that artists use to create their works, and help determine what is original to the work. By analyzing the materials used in artwork, conservation scientists can determine if they are consistent with the presumed date and location of origin of the piece. In some cases, microscopic and chemical examinations of paintings can be used to identify potential forgeries or authenticate works as being from a particular period.

Art conservation involves the interaction of many disciplines, including chemistry, archaeology, and history, and often entails more than a little detective work. Conservators must understand the period in which the piece was created in order to identify the materials used and thus determine the best way to preserve them for future generations. Artistic judgment can also be required, as, for example, when a piece has been modified over the years by many artists;

someone must decide what is the original. Conservation scientists maintain a balance between long-term research and ad hoc problem solving, with the ultimate goal of developing guidelines that will help art conservators choose safe and effective treatments for both modern and classic pieces.

Conservators and conservation scientists must work together to identify the materials used in a particular piece. This will affect how the object will react to different treatments. How will cleaning affect the long-term condition of the work? Is it stable enough to withstand transportation to an exhibition? How will the work stand up to future cleanings? What temperature, humidity, and other environmental conditions will best preserve the work in the long term? Many of these questions must be answered individually for every new piece; this process often requires using only nondestructive or micro-destructive methods of analysis.

In addition to how best to care for and preserve a particular piece, other types of questions may be asked. For example, from the mid-1800s to the mid-1900s, arsenic was used as a pesticide in the treatment of many wooden artifacts. Can these pieces now be safely cleaned and displayed, without risk to the conservator doing the treatment or to the patron visiting the exhibit? When artifacts are being returned to the native communities that created them, is the arsenic level or mechanism of transport likely to pose a risk to those handling the objects?

Scientists who end up working in museums arrive via a variety of paths. Some complete a traditional education in mainstream chemistry and then happen on the conservation field, often through a postdoctoral position or a job advertisement. Others had experience earlier or have an interest in art conservation and so pursue positions that relate to that field. In virtually all cases, their education is scientific, then they apply that background to the art field. Also, scientists sometimes train to be conservators, who are the ones who actually treat works of art.

Chemistry in the Creation of Art

The chemistry of pigments and coatings could fill a whole book—and in fact has filled many. Normally, these materials are used in large commercial applications, but there is also a market for artistic materials. Chemists today work to create nontoxic substitutes for older, more dangerous pigments, and new materials that perform and wear better. New materials also mean new properties—some pigments look different in different lights, and some change over time. Understanding and incorporating these differences into the art itself opens new vistas for the artist.

In fact, a semester-long Carnegie Mellon University course, called The Color of Minerals and Inorganic Pigments, in which half of the students are chemistry majors and half are art students, aims to teach chemistry students how their knowledge can be applied in the real world and to teach art students more about their materials. As the students work together, they understand more about what is needed and what is possible and why.

Profiles

Dana C. Lipp

Science Photographer
Dana Lipp Imaging

BS, Chemistry, Bridgewater State College, Bridgewater, Massachusetts, 1975
MS, Food Science, University of Massachusetts, Amherst, 1995

Current Position

As a freelance science photographer, Dana uses his knowledge and experience to accurately depict and communicate scientific principles and processes. He explains, "Well-constructed images can communicate complex phenomena without resorting to scientific terminology, an important aspect for those outside the discipline. While the technical skills of my job are readily apparent, marketing and solid business skills remain critical to long-term success for anyone running their own business."

Obviously, photography is one of Dana's main activities. He says, "My photographic assignments range from tabletop shooting of small objects, to shooting laboratories, manufacturing plants, instrumentation, and people. I might have to shoot objects in the studio or go out to visit a client to work directly at their site. Sometimes the assignments are broad, allowing me to use both my scientific and artistic eye; other times they are narrower, challenging me to work within their constraints."

However, the portion of time Dana actually spends shooting is not as large as one might think in respect to all his other activities. He notes, "It's typical for a small service business to have only 50–75% billable hours, and scientific photography is no exception." As with any small business, marketing is essential to getting assignments. Much of Dana's time is spent sending out promotional materials, exploring new leads, following up with potential clients, presenting his portfolio, and demonstrating the advantages of "insightful scientific images."

Dana notes, "Marketing and education take up a big part of my time. Most photographers or art directors really don't know the science behind technical subjects; this limits their creative vision, as demonstrated in the use of colored water, dry ice, and gels to enhance their photos and create visual interest. This science gap is especially unfortunate when the images are intended for scientists, who readily recognize a bogus image. It's my job to educate art directors through personal communication, articles, and workshops and explain the advantages of using engaging images with scientific depth, created by a photographer who understands science." Dana is always on the lookout for local scientific meetings as an opportunity to network and become better known, and he is working on educational tools (articles and workshops) to help increase awareness of science-art issues. He spends a little time each day reading and perusing publications to get a feel for who is using images, how they are being used, and what their strengths and weaknesses are to keep himself current on trends and techniques.

Dana occasionally commits some time to researching new subjects, trying out new ideas, and expanding his portfolio and the work he has available to sell as stock. A stock agency will review and approve a photographer's work then contract for a steady supply of images at some predetermined rate, on the order of 100 images per month, as determined by market needs. When a buyer contracts for use of an image from the agency, the photographer and the agency split the fees. Except for royalty-free images, a buyer actually pays to use the image for a specific period of time under specific conditions, and then it can be resold elsewhere.

Career Path

After graduating from college, Dana worked as a chemical technician at a small contract R&D company. He was able to work fairly independently initially, developing new assay methods for broad-based research projects. After the government funding for that project was reduced, Dana synthesized and tested polymeric sustained-release beads for subcutaneous implants. Although the work was rewarding and challenging, after three years it seemed time to move on and try something new. Dana recalls, "Little did I know then that my career path would ultimately take me full circle back to the small-company environment many years later."

Dana next obtained a position at Ventron, a company that manufactured metal hydrides and some biocides. There he continued providing analytical support for customers in the field, production, product development, and research. He says, "I was given a lot of challenging, non-routine work, which I really loved. Whenever someone found a precipitate, or an odd odor, color, or behavior, I was called in to investigate. It took creativity and problem solving, and I enjoyed talking to the developers and plant operators who had directly observed the problem." However, after 10 years there, he had seen most of the problems several times over and again sought new challenges.

Dana worked for a short time a small R&D company, but the instability and uncertainty of their business forced him to look elsewhere. He soon obtained a position as a senior chemist at Ocean Spray Cranberries, where he was able to apply his previous problem-solving experience to a new area—juice beverages. During his employment there, Dana took advantage of company-sponsored classes and obtained his master's degree in food science. Dana recalls, "It was a great opportunity to broaden my experience and education in this area, and I gained a newfound appreciation for the uncertainties inherent in the composition of food products. It was quite different from specialty chemicals, where I'd worked for most of my career."

After 10 years, the first downsizing in the company's history forced Dana to reexamine his career goals and his ideal work environment. His severance package included several months of career-counseling classes. Dana says that the personality tests they used reinforced what he already knew about his strengths and preferences." He notes, "While I didn't learn anything new or surprising, it gave me the confidence and the words to better articulate my skills. This was an exciting time, full of possibilities and newfound directions, but the pressures of a regular paycheck soon dawned, and I was back looking for full-time work again."

Dana used his network and found a position in the customer education department of the Waters Corporation. While there, he managed the training lab and occasionally trained customers. However, after further reflection on what he had learned from the career consultants, he came to realize that he would be happier out of the large corporate environment.

Dana applied for and was accepted into a highly selective, three-month intensive course in entrepreneurial training given by the Massachusetts Department of Employment, which was the final push he needed to start his own business. He was taught the important, practical aspects of running a business, and the training ended with each student's writing a business plan and presenting it to local bankers.

Dana recalls, "My business plan was based on my formal experience and skills in solving technical problems in chemistry, combined with my personal avocation, photography. For many years, I had been turned off by advertising that used implausible glassware setups filled with colored water and dry ice fog. I found them off-putting and, quite honestly, insulting. I saw one on the cover of a trade magazine and wrote to the editor. In my opinion, photos of this sort disengage the reader or viewer and devalue the products they represent, particularly to the scientists that view them. It concerned me that the images were not used to their fullest potential; those largely responsible understand only the visual and have little understanding for the science they are supposed to represent. So we end up with photographic clichés—the scientist holding a test tube up to the light in the eternal quest for truth and understanding. This was the cover photo for the chemistry set I had in the 1960s. We should expect better today!"

Dana spent the next few years marketing his skills as a scientific photographer through Dana Lipp Imaging, and doing adjunct teaching at Bridgewater State College. He says, "Adjunct teaching was very rewarding, but there was little money, lots of travel, and little security." Dana decided he needed a steady paycheck to supplement his freelance photography work.

For the last year, he has been working at pION, a 25-person company that performs testing and sells instrumentation to measure physicochemical constants and the absorption rate of drugs. As a product specialist, Dana is required to take on many tasks, including testing, making technical sales presentations, and installing equipment and conducting training at customer sites. He also does photography for customers' products, which helps his portfolio.

He still takes photographic assignments, which sometimes include performing product optimization, on his own time. He says, "I recently worked on an antioxidant salad dip and obtained eight hours' worth of time-lapse images showing treated and untreated controls. You can easily see the differences as the entire sequence plays back in only a minute."

As the owner of his own small business, Dana has to wear many hats —"customer relations, marketing, advertising, photography, competitive researcher, trainer, instructor, director." He explains, "There's certainly no shortage of variety! One of the biggest challenges is learning how to balance different priorities and skills, and knowing where your time is best spent!"

Dana also enjoys variety in the types of people he works with every day. He interacts with scientists, marketing people, company presidents, administrative assistants, and many others. Dana notes, "Occasionally, clients will want me to take images of the employees doing their jobs, or of the company president or management. You've got to set up, be ready for them, and work fast. Presidents won't sit still for much more than one minute, then they're off."

Dana is excited by the challenge of taking a subject, concept, or scientific principle and creating a powerful image within the constraints of budgets, time, and other resources. He enjoys the process of seeing that photography engenders—creating an image that is visually simple and engaging and that has scientific depth and accuracy.

On the other hand, he does not enjoy the precision required to get the images he visualizes to come out perfectly. He observes, "Oftentimes, many small adjustments are needed to get things just right in the scene, but the subjects don't always cooperate. Small things fall over, cords are a tripping hazard, I have to crouch in very uncomfortable positions behind the camera. It all looks easy until you try to do it!"

Unfortunately, for most science photographers, the income potential is not enough to support a full-time career. Overall, however, Dana is happy with the way his career has turned out. He says, "There were several years of trying to fit into the large corporate environment, but I did learn some good things along the way, and the sum total is who I am today. I feel very fortunate; my experiences have also helped me redefine what success means to me. It's something we all do, I think, as we get older.

"So now I feel it's taken me a long journey to finally find my niche, but it has been worth the wait!"

Advice

Dana says, "There really is no 'typical' path for a science photographer. Rochester Institute of Technology has a very well-respected program in photography, but it is more about the science of photography than about photographing science. Resources are very sparse, indeed! The few science photographers out there have a scientific–technical background. Some of them do it on a part-time basis, and a few are lucky enough to provide it as a service in academia, but that is very rare.

"It was important for me to learn as much I could about chemistry. To learn how to evaluate, think, and work scientifically—how to form a hypothesis, design an experiment to test it, and critically and objectively evaluate the results. These skills are useful everywhere, not just in science! This is the fundamental basis for problem solving, in which I've obtained my analytical expertise and which I am now applying to photography. I still find myself asking the questions, What works? What doesn't? Why doesn't it? What can I do better? Is this like anything else I already know? The scientist learns how to measure and observe; so must the photographer. There are a lot of skills common to both disciplines. Experiment, and allow yourself to make mistakes—plenty of them. Learn the rules, then break them; this is how breakthroughs happen. Experience is the best teacher. Work in industry; get some real-world experience. Learn to combine creativity with being analytical; both are essential to making successful images. And the most important advice of all? Have fun in whatever you do, and let your heart guide you.

"Resources are very sparse, indeed! You're pretty much on your own."

Predictions

Looking toward the future, Dana says, "Digital photography and image-processing software has enabled many people to do their own work. Despite the fact that image quality is widely variable, the reduced costs make it harder to persuade clients to hire a professional, especially in light of tight budgets.

"Digital photography and the Web will make images of all types more commonly available. Downsizing and reduced budgets will challenge businesses to do more with less. In order for a photographer be successful, potential clients must be able to see the extra value in the images science photographers create."

Sharon Davis Alderman

Weaver, Writer, Teacher
Sharon Alderman Handweaving

BS, Chemistry, Harvey Mudd College, 1963

Current Position

Sharon Alderman has been a full-time weaver since 1970. She teaches and lectures widely and in 1996 was awarded the Utah Governor's Award in the Arts. Her work has been shown in competitions and invitational exhibitions throughout North America and is held in private, corporate, city, county, state, and federal collections.

Sharon runs her own small weaving business. "I do everything here. I book teaching trips, write books—three so far—and magazine articles—well over 100, and design and create fabrics. I work every day, returning each morning to what I stepped away from the night before." Typical tasks for Sharon include measuring warps (the long threads on the loom); dressing the loom (putting warps on); threading the loom; filling shuttles; weaving, inspecting, and removing the cloth from the loom; making any necessary corrections; and then washing and pressing the finished fabric. She also handles correspondence, prepares for workshops, and writes magazine articles.

For 10 years, Sharon has woven prototype fabrics for various engineering firms. Among the exotic materials used in these prototypes are nylon monofilament, metal filaments, fusible materials, and nickel-plated carbon fibers. Sharon's scientific background makes her uniquely qualified to construct cloth from unusual materials.

She recalls, "In one project, the engineers were interested in an alloy that I handled as a fine wire, varying in width from .007 to .003 square inch cross section. The wire had a memory of sorts, becoming straight again when bent—unless crimped, and it was elastic as well. If pulled with force beyond a certain point, it stretched about 10%, and when the force was relaxed the wire resumed its original length. They were trying to find commercial applications for this wire. I wove with it, making cloth that used it in the weft, that is, crosswise, direction for uniforms to protect the wearer against flying metal—shrapnel, for example. I wove with it in both directions for use in making gloves for butchers to protect their fingers from sharp knives. I wove it into parachute fabric—very fine nylon woven very closely—for use in parachuting heavy things (such as jeeps or boxes of supplies). I wove it into fabrics that disappeared in use, except for the wire: one that would fuse non-woven textiles together, and another that dissolved in water for stents for arteries. I had to find the fusible material and figure out a way to use it so that the result would not be gooey; that is, I worked out a structure that would be extremely stable so that the number of wire filaments per inch in both direction was uniform and to the given specifications while using as little of the fusible material as possible. I had to locate the material that dissolved and likewise use it to make the wire spacing very reliable.

"In short, I was being asked to do things that I had never done before and at first thought seemed quite impossible. Nevertheless, I did them."

On another project, Sharon worked with nickel-plated carbon filaments to make cloth of different densities that would be very lightweight and yet would conduct electricity. It was to be used as the "skin" of an airplane that was invisible to radar, to cover the mast of America's Cup racing vessels, and for the frames of ultra-light bicycles. She says, "The problem here was that although the fibers had very high tensile strength, they had no shear strength. When woven on a handloom they had to be lifted to allow the crossing fibers to be woven in, which caused breakage."

Sharon has also "woven grill cloth for an acoustical engineer who makes speakers for sound systems that cost more than my house." Each new material brings new challenges in how it behaves during the weaving process and how the final material will behave once it's been woven.

Career Path

Sharon has always been interested in textiles. She grew up in a small, agricultural town in central California, where she made all her own clothes. She would imagine fabrics that she wanted to find but could not.

During her second year in college, a dorm mate showed a group of people a length of black-and-white tweedy wool fabric that she had woven. Sharon recalls, "I was knocked out. The next term another dorm mate showed us a Rya rug she had woven. I decided that someday I would learn to weave." However, college classes six days a week and labs five afternoons a week, plus two part-time jobs, left no time to learn weaving.

After college, Sharon spent the summer at Washington State University, where she expected to attend graduate school. Her plans changed, and she ended up in Bakersfield, California, working in a testing laboratory. She tested bacteria in water, the fat content of ground beef samples, the percentage of various components in cores from solid rocket fuel, the permeability of various membranes to hypergaulic liquid fuels for the Jet Propulsion Laboratory, and all sorts of other things.

After a time, she moved to Cornell, where she was a research assistant. Most of her work there was carried out in a glove box, "although the distillation of fuming HF was done just under a hood—for obvious reasons," she notes. After a couple years there, she and her husband wanted to start a family. Since her work was potentially harmful to a developing fetus, Sharon quit. After her daughter was born, Sharon stayed home to raise her. While doing this, she discovered a weaving class being offered nearby. She missed the first class but was able to take seven one-and-a-half-hour lessons.

Sharon recalls, "Once I started weaving, I wanted to know more and more. Reading everything I could find on weaving, design, and color helped a lot, but the Basic Design course at the University of Utah Graduate School of Architecture made a big difference."

When she decided to start her own business, Sharon applied for the Certificate of Excellence in Handweaving sponsored by the Handweavers Guild of America in 1976, the year it was introduced. She says, "I realized that I needed some sort of credentials to teach weaving, and this certification launched both my teaching and writing."

Most days, Sharon works alone in her studio, but she interacts with workshop organizers and workshop participants, and with yarn and equipment suppliers.

She notes, "To be a designer of woven fabrics one must understand and be able to use weave structures, know how choice of fibers and yarn styles will affect the cloth, understand how to used color effectively, and know how to finish cloth. In addition, I must have a good vocabulary and the ability to transmit information—orally and through writing—to others in a clear, easily understood manner.

"The best part of my career is the joy of having made something beautiful that will function as it is asked to do and that would not have existed if I hadn't made it. The creation of weave structures I have never seen before is also fun. The worst part is the financial instability and dismal compensation."

Sharon's chemistry training gave her the discipline required to stay on task and see a project through to completion. Being able to complete projects, even when one's interest has waned, is crucial for an independent artisan.

Sharon says, "I have done a fair amount of work over the years preparing prototypes for engineers. They need fabrics to test to see if they have the required properties. To produce these fabrics I must devise ways of working with exotic materials. Because of my technical background I understand the engineers' language and what they need. This work is quite difficult but very interesting.

"I think the work with the 'memory fiber' was the most challenging and the most fun. I love figuring out how to do something that I thought was impossible. The cost of the alloy they were interested in was very high, so I had to make the cloth 'work' the first time out. I did a lot of thinking before I started each piece of work and sometimes wove with a less valuable, but similar, material to be sure I understood how it would behave. I discovered that introducing twist into the filament caused it to kink and fail, so I devised a way to weave without twisting it. When used in the warp and in the weft there were problems. The filament would curve if the diameter of the curve was at least an inch, but I didn't have that option when weaving a fabric with 100 threads to the inch in the warp direction. When used in just one direction—along with other threads—the filament forced the threads that crossed it—whether in the warp or in the weft—to do all the curving, while they remained straight. When used to cross itself, making a cloth of the required fineness was a challenge and forced me to change the way I put the weft across the warp and how I beat it into place.

"I think to be able to do the things I did for the various engineers with whom I work it takes a special way of approaching a problem. One needs to define exactly what is wanted and then figure a way to get there. It sounds very straightforward but requires a deep understanding of loom function, loom mechanics, weave structures, and material properties, and a lot of personal tenacity. The last quality is probably most important! I have always loved puzzles, so this work suits me very well."

Advice

Sharon says, "The fact that I am a weaver who weaves for engineers is so completely unlikely that I can't imagine anyone doing this on purpose. While I love what I do, I have none of the financial security I once envisioned for myself.

"But, for the individual who is really motivated, my advice would be to get to know weaving and the way that looms work. Understand weave structures and work to connect all the things you know, whether you learned them in a chemical or engineering or weaving context."

Predictions

"I think that fabrics can be made that will do a number of things our ancestors never dreamed possible," Sharon says, adding, "They can be made so that they could keep us warm—not simply by insulating us—or cool and can monitor temperature, pulse rate, and possibly even blood pressure. I can imagine fabrics made with fiber-optical materials for very decorative effects—a pile with light coming out of the cut ends, for example. I think it would be fun to dream up and bring these things into being."

Alison Murray

Conservation Scientist and Associate Professor, Art Conservation Program, Department of Art
Queen's University, Kingston, Ontario, Canada

BS, Honors Chemistry, McGill University, 1987
MS, Materials Science in Engineering, Johns Hopkins University, 1990
PhD, Conservation Science, Johns Hopkins University, 1993

Current Position

Alison, the only scientist in the art department at Queen's University, is an associate professor in Canada's only master's program for art conservation. Most of her students learn how to be conservators, and they study in one of the following three streams: paintings, paper, or objects. There is also a smaller stream in conservation science, where students with an undergraduate degree in science or engineering can do conservation-science research. Alison teaches the science component of the program, which includes classes in polarized-light microscopy, other instrumental methods of analysis used in the field, and properties of materials (structure and degradation), and a science research project.

She also conducts research into the degradation of art objects using microscopy, imaging, and other nondestructive analytical techniques, and works to identify artists' materials and techniques. Currently, she is conducting a research program aimed at optimizing the cleaning treatments used in the conservation of acrylic paints and grounds that integrates information from mechanical, chemical, and surface analysis. She has worked with people from a variety of academic disciplines—including conservation, conservation science, art history, physics, chemistry, and engineering—as well as a prominent manufacturer of artists' materials.

As a scientist in the art department, she studies the effects of various cleaning solutions on acrylic paints. There has been a lot of work done on oil paints, but acrylics were first used by a great number of artists in the 1960s, so not as much is known about how they change as they age and how they can be cleaned without damage.

Alison spends about 40% of her time teaching, 40% on research, and 20% on administration, which varies throughout the year and from year to year.

Career Path

Alison always loved art and visiting galleries, but she found herself becoming more and more interested in science in high school. When she went to McGill University, she pursued a degree in chemistry.

During her undergraduate studies, Alison went to an extra physics tutorial and found, to her surprise, that she was the only student there. She ended up talking to the professor about his research with thermoluminescence dating of artistic works, and about the conservation-science work being conducted by the Canadian Conservation Institute (CCI) and Parks Canada. Alison was intrigued and very excited, went to visit their labs, and was able to get a summer job at CCI. She recalls, "I was lucky to hear about the field in the first place. When I visited CCI, the scientists said there were never any summer jobs. Suddenly one opened up—I competed and was fortunate enough to get it!"

When she finished her undergraduate degree, Alison found a brand-new graduate program where she could complete her PhD in a materials science and engineering, in a program offered jointly by the Johns Hopkins University and the Smithsonian Institution. She says, "I had three supervisors during my graduate career—one from Hopkins, one from the Smithsonian, and one from the National Institute of Standards and Technology (NIST)."

After obtaining her PhD, Alison "wrote to everyone," and found a fellowship in the Analytical Research Services Division at the Canadian Conservation Institute, where she used a variety of analytical techniques to examine museum objects, artists' materials, and conservation materials in support of long-term research projects and service requests.

Alison then received a fellowship from the Samuel H. Kress Foundation to work in the Scientific Department at the National Gallery in London for four months, investigating methods to image environmentally induced defects in paintings.

She moved back to Canada and obtained a contract position at the National Research Council (NRC) of Canada, doing polymer research on roofing materials. She recalls, "I thought this might lead to a postdoctoral position, perhaps. And they do work on historic buildings, so I thought there might be some job prospects in the field at NRC." However, shortly after she started at the NRC, Queen's University asked her to interview. They offered her a position, and she started there a month later as an assistant professor.

Alison very much enjoys the mix of science, engineering, and the humanities that her work requires. She works with a community of scientists from around the world who are interested in solving difficult problems. Included are organic, inorganic, physical, polymer, and analytical chemists; physicists; geologists; biologists; engineers (for example, mechanical engineers); and non-scientists (including conservators, art historians, archaeologists, and artists).

Alison looks forward to continuing her international collaborations and would like to welcome more students from the sciences, engineering, and conservation; many students of conservation major in the humanities. She finds it enormously satisfying to organize conference sessions where colleagues from around the word share their expertise.

Alison feels that she is very fortunate to be able to work on cultural materials and to be able to see behind the scenes at museums and galleries. She notes, "This is a very applied field, where conservators need quick answers to very specific questions. The challenge is that there has been limited research and little funding for research. At the heart of it all, chemistry is key to understanding the properties and degradation of materials found in art objects."

Advice

Alison says, "Even though there are limited jobs and opportunities for experience, if you are really interested, get the right background and practical experience, and your persistence is likely to pay off. Things have worked out for me in ways I could never have anticipated."

Predictions

"I hope that more professors in mainstream science do some research in this field. There are many difficult questions still to be answered by people from a number of fields," Alison notes.

ADDITIONAL RESOURCES

Alderman, Sharon. 2004. *Mastering Weave Structures: Transforming Ideas into Great Cloth.* Loveland, CO: Interweave Press.

The American Institute for Conservation (AIC) (aic.stanford.edu) and their specialty group on Research and Technical Studies (RATS) (aic.stanford.edu/sg/rats/about.html) provide useful online resources for conservation professionals

The Canadian Association for Conservation (www.cac-accr.ca)

The Canadian Conservation Institute (www.cci-icc.gc.ca)

Conservation OnLine (palimpsest.stanford.edu) provides resources for conservation professionals

Ember, Lois. Chemistry and Art. 2001. *Chemical and Engineering News 79.* (July 30)79(31): 51–59.

Frankel, Felice. 2004. *Envisioning Science: The Design and Craft of the Science Image.* Boston: MIT Press.

International Centre for the Study of the Preservation and Restoration of Cultural Property (www.iccrom.org)

The International Institute for Conservation (www.iiconservation.org)

Visions of Science Photographic Awards (www.visions-of-science.co.uk) is a yearly science-photography competition in the United Kingdom

Conclusions

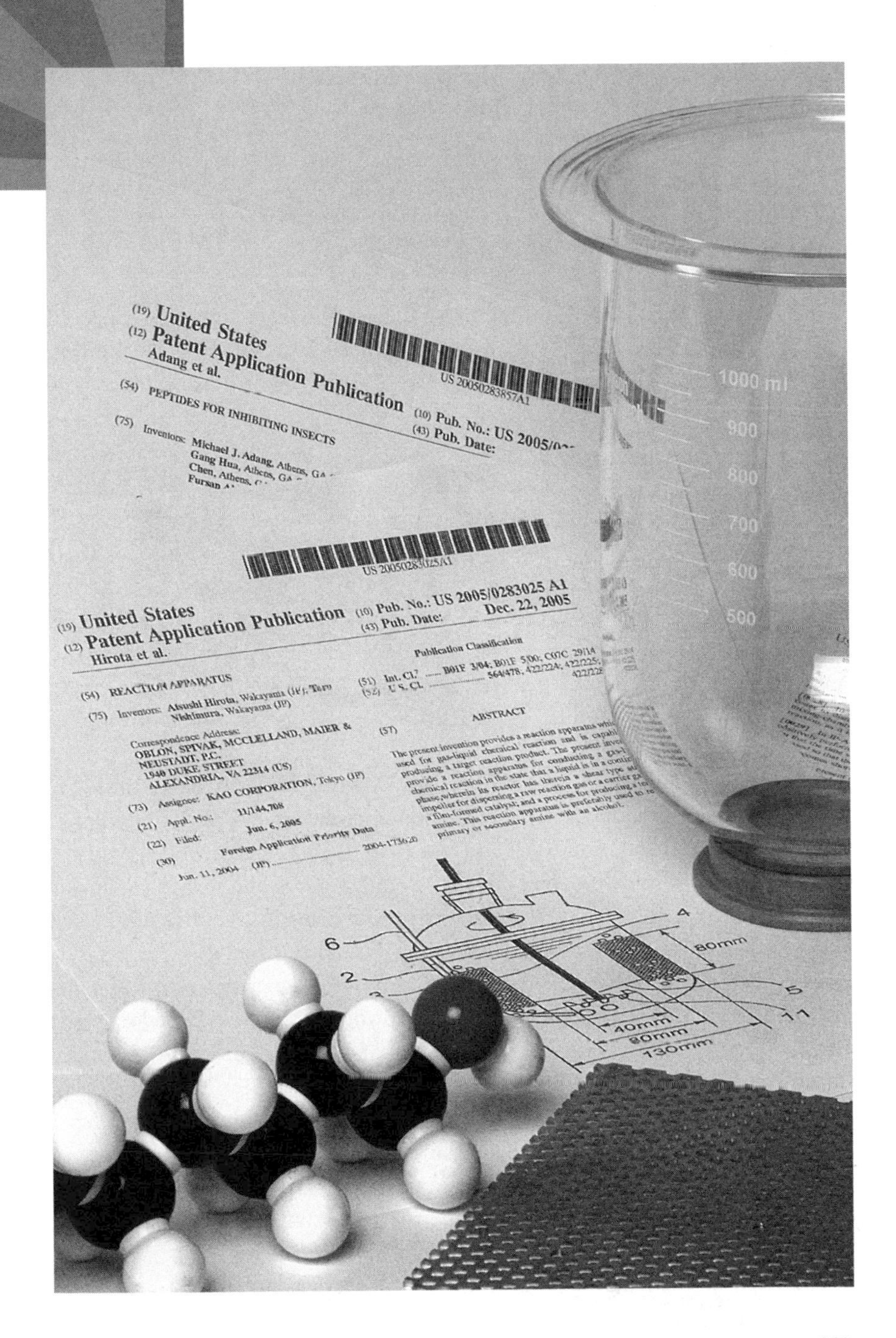

BY NOW I hope you realize that this is not the end but just the beginning of your career exploration. The preceding chapters have revealed some new professional possibilities—and maybe you found one or two exciting enough that you want to learn more about them. This chapter will provide insight on how to do that.

As chemists, we are taught how to solve problems and how to find and critically evaluate data. We start by doing background research, identifying a problem, conducting an experiment, then analyzing the results and applying what we learned to the next question. You can apply those same steps to the question of what career path will best suit your personal skills, knowledge, interests, and values.

Redirecting your career path requires a thorough understanding of your own motivations, and an understanding of what a particular job or field requires. You need to figure out not only what is available but what you want. Ask yourself questions such as: What kinds of problems do I solve well? What do I most enjoy talking about? What do other people say I do well? What characteristics truly define me? What accomplishments am I most proud of? Browse library's collection of career titles and see what jumps out at you.

Chemists usually have strong analytical skills, but sometimes focus their thinking on the most probable solution and don't consider all the outlying possibilities. Be aware of this tendency, and make sure not to discount options too early. There are shades, ranges, and sides to every job, and you may come to the best answer in a roundabout way. Once you have a hypothesis for a career option you think is a good fit for you, learn as much as you can about it, and find a way to try it out.

Remember also that your professional abilities and needs will change throughout your career, so you will need to continually reevaluate how well your current situation matches your ideal and make adjustments accordingly.

Like most things in life, the only way to really understand what a particular job is like is to experience it firsthand. Especially those just entering the work world should seek out positions that let them try many different types of tasks. Seek out internships, co-op programs, and summer and part-time work as valuable sources of real-world experience. Later in your career, you can take on additional responsibilities at your current company to move in a new direction. At any career stage, use volunteer opportunities to learn new skills and meet people. As an added benefit, you'll have concrete accomplishments that demonstrate your abilities to potential new employers. You may learn that you love the field and want to pursue it full-time, or you may learn that you need to move in another direction. Reflect on past job experiences, and figure out what specific aspects you did and did not enjoy. "Experience can be a wonderful teacher, if you let it," says Richard Bretz, an ACS career services presenter.

The second-best way to understand something is to talk to people who have done it. Use your network! Ask questions of people already working in jobs you find intriguing, and find out what they do on a daily basis. Discover what types of tasks they spend most of their time on, what types of people they interact with, and what types of things they need to know. Job titles are not enough: you want to find out what the person really does every day. What do they like and not like about it? What brings them their greatest personal satisfaction at the end of the day? Attend meetings of their professional societies, and see what kind of people work in this field and what kinds of things they do.

Find out not only what skills and knowledge are required but what personality traits are needed to be successful in that field. Should you be outgoing, self-motivated, and love to talk to people? Or does it require a lot of solitary work and meticulous attention to detail?

Investigate not only what you'd like to do but where you'd like to do it. Currently, small companies (those with fewer than 500 employees) hire more chemists than large companies, and that trend is expected to continue. Small companies may be harder to find, but they offer tremendous opportunities for teamwork across disciplines, in addition to opportunities to take on new and nontraditional tasks. With fewer people to do the work, everyone has to do a little of everything.

Always be on the lookout for opportunities to try something new—seek them out, and consider them seriously when they are offered, especially if the risk is low. It's never too late to try something new, and in the worst case you'll find something you don't want to do again.

In addition to learning about what you enjoy and what professional opportunities are available, keep an eye on trends, and consider what opportunities they might bring. The American Chemical Society's report *The Chemistry Enterprise in 2015* says that "to maintain lifelong success, chemists will need to develop, maintain and grow skills through continuing education, both formal and informal. In a world of commoditization, personal differentiation and marketing will be critical." Knowing what is coming up will allow you to prepare for the needs of the new market.

On the other hand, don't go into a field just because you hear there are lots of positions available—they may not be there by the time you get done with your training, and you won't be happy if you spend all day doing something you don't enjoy. Think broadly in terms of what you can do—you are not defined by your degree but by the sum of the experiences, skills, and knowledge you have accumulated. Your talents can be applied to a variety of different fields, and you can (and should!) always keep learning new skills and adding new areas of expertise.

This is not a process that can be rushed. If you admit you're discontented and start investigating your options for something better, you're already a step ahead of those who just accept where they are. Changing your career direction requires soul-searching, research, planning, and courage. But in the end, getting to spend your time doing something you love is worth it.

Chemistry is everywhere, and your perfect career is out there—you just have to find it.

Good luck!

ADDITIONAL RESOURCES

Books

Bolles, Richard N. 2006. *What Color Is Your Parachute?* Berkeley, CA: Ten Speed Press. General advice and worksheets to help job-seekers figure out what they do well, what they enjoy doing, and how they might find a job doing it.

Borchardt, John K. 2000. *Career Management for Scientists and Engineers.* New York: Oxford University Press. Not a "how to find a job" book, this is a "how to manage your career" book.

Brown, Sheldon S. 2000. *Opportunities in Biotechnology Careers.* Chicago, IL: VGM Career Books.

Camenson, Blythe. 2001. *Opportunities in Forensic Science Careers.* Also see other books in the series published by VGM Career Books.

Friary, Richard. 2000. *Job$ in the Drug Industry.* New York: Academic Press. Discusses the types of jobs available in the pharmaceutical industry, why they are desirable, and how to go about finding suitable openings.

Harkness, Helen. 1999. *Don't Stop the Career Clock: Rejecting the Myths of Aging for a New Way to Work in the 21st Century.* Mountain View, CA: Davies-Black Publishing.

Owens, Fred, Roger Uhler, and Corinne Marasco. 1997. *Careers for Chemists, A World Outside the Lab.*, Washington, DC: American Chemical Society Press.

Robbins-Roth, Cynthia. 2005. *Alternative Careers in Science: Leaving the Ivory Tower,* 2nd ed., New York: Academic Press.

Rodman, Dorothy, Donald D. Bly, Fred Owens, and Ann-Claire Anderson. 1995. *Career Transitions for Chemists.* Washington, DC: American Chemical Society Press.

Rosen, Stephen, and Celia Paul. 1998. *Career Renewal.* New York: Academic Press. Career changing tips, and many stories about scientists and technical people who moved into new fields.

Stonier, Peter D. 2003. *Careers with the Pharmaceutical Industry,* 2nd ed. West Essex, UK: John Wiley and Sons. Potential opportunities in the United Kingdom's pharmaceutical industry.

United States Department of Labor. 2006–2007. *Occupational Outlook Handbook.* Descriptions of all sorts of jobs, mainly traditional, including working conditions, training required, outlook, and earnings. Good place to look for ideas.

Woodburn, John H. 2002. *Opportunities in Chemistry Careers.* New York: VGM Career Books.

Young Kreeger, Karen. 1998. *Guide to Nontraditional Careers in Science.* Philadelphia, PA: Taylor and Francis.

Web Sites

American Chemical Society (www.chemistry.org/careers) posts resumes, interviewing and specific job-hunting advice, salary and employment-outlook data, and personalized career-consulting for members.

Science's Next Wave (nextwave.sciencemag.org) provides career-development information and ideas for young scientists. It has a huge collection of articles by scientists in hundreds of careers and positions and is a great place to go to browse for ideas, or to get an insider's scoop on what a job in a particular field is really like.

Sloan Career Cornerstone Center (www.careercornerstone.org) is a resource center for those interested in careers in science, technology, engineering, and mathematics and includes general information and resources for pre-college students, college students, and early-career professionals. It covers aerospace engineering, bioengineering, biology, chemical engineering, chemistry, civil engineering, electrical engineering and computer science, engineering technology, geosciences, industrial engineering, information technology, materials science and engineering, mathematics, mechanical engineering, nuclear engineering, and physics.

TinyTechJobs (www.tinytechjobs.com) posts news, career information, and jobs in nanotechnology, microtechnology, biotechnology, and information technology.

Index